To Mum
With Lots of Love
from
Douglas and Suzanne xxx
August 84

D1783958

ALL THE BEASTS OF THE FIELD

All the Beasts of the Field

SYLVIA FENTON

Illustrated by David Astin

London
GEORGE ALLEN & UNWIN
Boston   Sydney

George Allen & Unwin (Publishers) Ltd,
40 Museum Street, London WC1A 1LU, UK

George Allen & Unwin (Publishers) Ltd,
Park Lane, Hemel Hempstead, Herts HP2 4TE, UK

Allen & Unwin Inc.,
9 Winchester Terrace, Winchester, Mass 01890, USA

George Allen & Unwin Australia Pty Ltd
8 Napier Street, North Sydney, NSW 2060, Australia

First published in 1984

British Library Cataloguing in Publication Data

Fenton, Sylvia
   All the beasts of the field.
1. Animal culture
I. Title
636.08'3   SF61
ISBN 0–04–925022–1

Set in 10 on 12 point Palatino by
Nene Phototypesetters Ltd, Northampton
and printed in Great Britain
by Mackays of Chatham

To Zena

CHAPTER 1

'Gotcha!' I yelled, and made a grab for him. He did a bit of fancy footwork, caught me off balance and sent me hurtling head-first into the ditch. Gibbering with rage and frustration, I wiped the mud out of my eyes and watched his fat, smug, fast-disappearing backside with murder in my heart. I love donkeys, and I particularly love Humphrey, but if he had been within grabbing distance at that moment I think I would have strangled him with my bare hands.

I had been chasing him for over an hour, but really it was no more than a gesture. I knew, and he knew, that I had no hope of catching him. He held all the aces. For one thing, he had four legs to my two; also he was younger and, being a non-smoker, a whole lot fitter. And he regarded the whole thing as a game. After all, he wasn't the one who would have to go and apologise to the neighbouring farmers for the devastation to their crops and stock.

By now there was no sign of him. From experience, I knew he would be either in Joe Sheppey's field, creating mayhem among his cows, or in Graham's barn eating his winter store of hay, or . . . Suddenly I knew with heart-lurching certainty that this time he would not be in any of the old familiar places. This time he would be favouring Julia with his company. For over a week I had

1

been trying to blot out the memory of the last time Julia had played reluctant host to Humphrey at the Big House, but now it all came flooding back.

The blacksmith calls at the Big House once a month, and owners in the vicinity take their horses and ponies along to have their hooves attended to. Humphrey's hooves had needed trimming, so it was arranged that I would take him along and the blacksmith would fit him in when he had attended to his other, more distinguished clients. What could be simpler or more straightforward? What indeed. To start with, Humphrey took one look at the head collar and immediately went into his 'They've-come-to-lynch-me' routine. Put his head in a noose? No way. Next thing he knew, he'd be strung from the nearest tree. A good forty minutes, half a pound of sugar lumps, three carrots and a tube of mints later (and to our mutual surprise), the halter was on. Well, almost. It was back to front and hung drunkenly over one ear, but this was no time for aesthetic considerations. We were already more than half an hour late. I was exhausted and in no mood for any more nonsense. I opened the gate and nudged Humphrey towards it but, for a donkey who spent most of his time squeezing out of mouse-sized gaps in the hedge, he showed a remarkable reluctance to leave the paddock when given the opportunity. He looked at me in horror: You want me to go through there? To leave the safety and security of my lovely home? To face the unknown dangers of the Outside World? You must be mad. He dug in his heels and refused to budge. I begged and pleaded and pushed and shoved. Nothing. Bribery was no good – I'd already squeezed that ploy to the limit, trying to get his halter on. Suddenly inspiration struck. I dropped the rope, walked out through the gate and set off down the path. Sure enough, before I had covered ten yards, I heard the clippety-clop of little hooves coming up behind me, followed by a 'hummpph'-ing in my ear. I took no notice but just kept on going, and Humphrey trotted happily

2

along behind me as though the whole thing had been his idea in the first place.

If I had thought that was the end of my troubles, I couldn't have been more mistaken. When we arrived at the Big House, my delinquent donkey was entranced. All those lovely people to goose and all those fancy horses with their fat backsides just begging for a friendly nip. Dogs to chase and cats to frighten the living daylights out of. This was fun! Why hadn't we done it before? The elegant, exquisitely jodhpured ladies took his boisterous behaviour in their stride with not so much as a raised eyebrow. For my part, I was desperately embarrassed and ashamed. I watched him careering madly over Julia's immaculate front garden, his hooves making dirty great potholes in her beautifully mani-cured lawn, and prayed fervently for death. He was eventually rounded up (not by me, I'm ashamed to say – I was too busy pretending I'd never seen him before) and tied to a tree which, with studied nonchalance, he proceeded to eat. I, with equal nonchalance, looked the other way and acted as though it wasn't happening.

The other horses had been watching this dreadful performance impassively, not allowing so much as a hint of the disapproval they undoubtedly felt to show on their aristocratic faces. No doubt about it, breeding tells. I was completely bowled over by their beautiful man-ners. 'Next, please,' said the blacksmith, and up they minced. 'Front hoof, please,' and an elegant hoof would be proffered. It had all been so civilised. I won't go into details about what happened when it came to Hum-phrey's turn. All I will say is that it took five people to hold him, the blacksmith suffered a severe kick on the shin and three horse-owners had their expensively shod toes crushed by several hundredweight of donkey. My humiliation was complete.

Now, the idea of calling on Julia and asking, 'Please may I have my donkey back?' appalled me. I just couldn't face it – not yet. I crawled onto the bank, lit a soggy cigarette and asked myself, not for the first time, what in the name of Heaven I was doing here anyway. Why was I chasing bloody-minded donkeys across sloshy fields, being lacerated by brambles, stung by nettles and bitten by insects I hadn't even known existed in my palmier days? Why was I living like a social outcast in the middle of nowhere, in a caravan with no electricity, no heating, no hot water? There wasn't even a proper road, so even if anyone wanted to visit me (which, of course, nobody did), they wouldn't be able to. My only companions were five cats, two toads and a donkey who valued my company so highly that he spent all his waking hours devising ever more ingenious ways to escape. I thought longingly of my nice, convenient, centrally heated London flat. I reflected on life in the Metropolis – the bustle and activity, the crowds, the noise, the pressures. That brought me back to reality with a jolt, because I knew there was no power on earth that would induce me to go back to all that.

My love affair with the country began when I was evacuated from the East End of London to Somerset during the war. Of course I had been on family trips to the 'country' – to Epping Forest and Hampstead Heath and Kew Gardens – but this was something entirely different. For one thing, you didn't have to embark on an expedition involving complicated bus and train journeys to get to the countryside here – it was all around you. For another, there was so much of it. It seemed to go on for ever. And it was so peaceful, with no noisy picnickers to drown out the sound of bees humming and birds singing. Best of all, it was full of animals: cows and sheep and pigs and chickens and goodness only knew what. To someone coming from the noise and grime and greyness of London, the peace and sense of space, the sheer beauty of the countryside was breathtaking. If anyone had told me at that time that I had been evacuated to Paradise, I would have seen no reason to disbelieve them. I was particularly impressed by the obvious honesty of country people. Nuts and fruit and flowers and vegetables were all growing out in the open, and nobody nicked them! They wouldn't have lasted five minutes where I came from. Our next-door neighbours in London had a stunted apple tree which produced about half a dozen wizened apples every year. They never had a chance to eat them, for long before the fruit was ripe, swarms of schoolchildren would climb over the garden wall and strip the tree of the walnut-sized, green, totally uneatable apples.

At the end of the war I came back to London and, after a series of jobs, began a reasonably successful career in advertising research. But I never lost sight of my dream that one day I would live in a cottage in the country. Common sense told me that, short of winning the football pools (which was extremely unlikely, as I didn't do them), there was not much chance of realising my dream until after I had retired. By then I would be too old to enjoy it and, with the way property prices were soaring, unable to afford it anyway.

Then I woke up one morning and the sun was shining. I thought of the traffic-choked drive into the centre of London, the stuffy office, the crowded streets and surly people and I said, 'The hell with it – do it NOW.'

Of course, it wasn't quite as simple as that. First I had to find my cottage. I knew exactly what I wanted: an isolated cottage, condition immaterial, with at least an acre of land and at a price I could afford. Unfortunately, several thousand other people had the same idea. This was at the height of the property boom, when derelict shacks with a few square yards of 'grounds' were being snapped up immediately they came onto the market, given the full cottage treatment (exposed fibre-glass 'beams', bijou ingle-nooks and pseudo-Tudor façades) and sold again at astronomical prices.

During the next three years I must have looked at about a thousand 'desirable properties' (desirable to

whom, I wondered) without seeing anything that made me feel, 'Yes, I could quite happily spend the rest of my life here.' But once I learned to cut through the estate agents' guff, things were much easier. 'Semi-rural', I discovered, meant suburbia. 'Much sought-after' meant over-priced. 'Scope for improvement' = falling apart. 'In need of some attention' = fallen apart. 'Unspoilt' = no gas or electricity. 'Bijou' = tasteless. 'Imposing' = impossible to heat. 'Compact' = Wendy house.

To be fair, some agents were painfully honest. For example, the property 'in quiet surroundings, with no noisy neighbours' turned out to be a de-consecrated chapel in a disused graveyard. You can't get much quieter neighbours than that! At the other end of the scale, there was the agent who sent me details of a 'delightful property in peaceful rural surroundings'. When I asked directions at the local post office I was told, 'Keep straight on along this road till you come to the cement works, turn right and it's just past the brewery.' I didn't bother to pursue that one! In time I learned that the best way to find out about any hidden snags was to have a quiet drink and a chat with the landlord of the local pub. That way I discovered that the 'quaint little cottage nestling in a hollow' was so situated that the effluent so casually flushed away in the morning was quite likely to come creeping back into the kitchen in the afternoon, that the 'charming dwelling set in two acres' was scheduled to have a dual carriageway bisecting the garden within the next five years, that 'Chestnut Cot-tage' was downwind of a pig farm and that 'The Willows' was built over an old rubbish-tip, into which it was rapidly sinking.

But I wasn't discouraged. I was convinced that some-where there was a cottage that was absolutely right, and that as soon as I saw it I would know, without any shadow of doubt, that this was the one. That's just how it turned out to be. An agency sent me details of a cottage so isolated that even they couldn't tell me exactly how to

get there, and in fact it took me two days to find it. At first sight it was pretty unpromising: completely derelict, with no means of access except for a footpath across the fields (no wonder I couldn't find it), no main drainage and no electricity. Water was piped from a neighbouring farm, but the cottage had been unoccupied for so long that nobody could remember where the outlet was. But it was beautifully situated, on top of a hill with unobstructed views in every direction. There were three acres of land, including a paddock which backed onto woodland. It was perfect. It was also sheer lunacy, but I had to have it.

Being a coward at heart, I waited until contracts had been exchanged before telling anyone about my projected move to the country. If I'd announced that I was planning to pitch a tent in Outer Mongolia, the reaction couldn't have been more dramatic – or more predictable. My friends were appalled. 'Nobody *lives* in the country,' they said. The country was for visiting or having picnics

in or passing through en route to somewhere else. It was for sheep and pigs and cows, not for people. As the closest most of them had ever come to the rural scene was leafing through *Country Life* at the hairdresser's, I wasn't too surprised at their reaction.

There was one person who was genuinely delighted at the news. I had met Jim Barry when he was an RSPCA inspector based in London and, even though he had since been transferred to Wales as a chief inspector, we still kept in touch. I had 'homed' a number of animals for him over the years, mostly cats and birds because my limited accommodation ruled out anything too large or demanding. But now, with three acres and no complaining neighbours, the sky was the limit!

I remember the very first waif I took in from Jim at my London flat. It was an African Grey parrot, although I only had Jim's word for this. At first sight it looked exactly like an oven-ready chicken, completely naked except for a tuft of drooping red feathers on its rump. Jim told me that the parrot, Archie, had been the pet of a police inspector and his wife. The wife had recently died and the police inspector found the parrot's constant repetition of phrases and expressions habitually used by her too distressing, so he had asked Jim to find a new home for the bird. Archie for his part had found the trauma of losing first his mistress and then his home too much for him, and had systematically plucked himself, feather by feather.

I could quite see the police inspector's point as Archie chattered incessantly, alternating between a high-pitched woman's voice and a deep masculine one. A typical dialogue would go something like this:

High-Pitched Woman's Voice: Reg? Is that you, Reg? Is it you, Reg, is it?

Deep Masculine Voice: Elsie? It's me, Elsie. It's me, Reg. Elsie?

HPWV: Want a cup of tea, Reg? Cup of tea? Tea, Reg? (Short pause)

DMV: Walkies, Brucie. Coming walkies, Brucie? Come
  on, Brucie. (Followed by a shrill whistle)
HPWV: Going to the pub? Going to the pub, Reg? Don't
  be long, Reg, supper's ready. Going to the pub?

Archie spent his days sitting on top of his cage. Every
time I passed, he would give a long, low wolf-whistle
and say, 'Hallo, pretty boy!' Half of me was flattered, the
other half was a bit worried! At night I put him back into
his cage, and this was when the trouble started. He
didn't want to go. He stood defiantly on top of the cage,
chest out-thrust and wings akimbo, daring me to catch
him: 'Come on then, pretty boy, come and get me.' As
soon as I made a move to grab him, he would be off with
a squawk to the other end of the room. We went through
this ridiculous pantomime every night. I knew I had no
chance of catching him and that my only hope was to tire
him out. Usually, after about ten minutes, he conceded
defeat and came quietly, but one night he dug in his
heels and resolutely refused to give up. After half an
hour he was still giving me the run-around and it was
quite obvious that I was going to tire long before he did.
Exasperated, I yelled, 'Will you come here, you miserable
old crow!' He turned round, fixed me with a long, hard
stare and said, 'Piss off.' I was shocked to the core – and
found myself wondering about that highly respectable
couple, Reg and Elsie!

Excitedly, I told Jim about the cottage – the isolation,
the woods, the paddock. 'Paddock, eh?' His eyes lit up.
'I've got just the thing for you. How would you like a nice
little donkey?'

A donkey! My very own donkey! I couldn't believe it. I
have always been mad about donkeys, ever since my
first donkey-ride on the beach at Margate. To think that I
was now to have one of these sweet, gentle creatures of
my own. Of course, I was incredibly naïve at that time
and didn't realise that there are donkeys and donkeys –
or rather, donkeys and Humphrey. Jim told me that the
donkey had been found in a waterlogged field in the

9

depths of winter with no drinking water, food or shelter. His ears had been badly infested with parasites and his hooves dreadfully overgrown, and all in all he was a very sad little donkey indeed. However, with expert care from Jim and lots of love from Jim's family, he was making very good progress and now the main requirement was a secure, loving home.

In fact, right now this was my main requirement too. I had sold my flat and the cottage in its present state was quite uninhabitable. The obvious solution was to buy a second-hand caravan and park it in the garden. That way I could keep an eye on progress as the builders got the cottage ready for me to move in: naïvely, I assumed that this wouldn't take more than nine months at the most. It was agreed that Jim would keep the donkey until I was sufficiently settled to take in waifs and strays.

CHAPTER 2

One of the things that worried me most about the move to the country was how the cats would adapt to the change. Not only would they be leaving the only home they had lived in since kittenhood, they would face the additional trauma of being plunged into a totally alien environment. How would they cope with the vastness of the countryside after a lifetime in a London flat? What if they developed galloping agoraphobia and refused to set paw outdoors? When they did venture out, they would be exposed to dangers they had never encountered before – foxes, stoats, traps, not to mention gun-happy maniacs shooting anything that moved. And snakes: the woods were probably alive with snakes. How could they be expected to know the difference between a venomous adder and the battered rubber snake they had played with since babyhood? Fast losing my grip on reality, I agonised over the dangers of exposing town-bred cats to country ills such as foot and mouth, fowl pest, swine fever, foot-rot . . . 'Oh, don't be ridiculous,' I told myself.

Of the five cats, the one I was most worried about was Rufus. He was an elderly gent of fifteen and, like most people in their twilight years, he liked a well-ordered, totally predictable life. He insisted on fish for breakfast and liver for supper – every day. If he was offered liver for breakfast or fish for supper, he was most put out. And

if, Heaven forbid, neither fish nor liver was available and he was given tinned cat food, he was grossly offended. Not only would he not eat it, he would go to great pains to be seen to be not eating it. He sat by his dish, looking pained. From time to time he shot looks of extreme distaste at it. In his more inspired moments I was treated to his Oscar-winning 'I-am-wasting-away-but-I-will-starve-before-I-sacrifice-my-principles' performance. As he was a very large marmalade and white half-Persian weighing about twelve pounds, his histrionics cut very little ice with me.

For the first thirteen years of his life, Rufus was an only cat. So when Min, a microscopic tortoiseshell waif, came into our lives, I was a bit apprehensive about how Rufus would react. I needn't have worried because he simply pretended that she wasn't there, to the extent that, if she happened to be on his chair, he just sat on top of her. Min was about six months old when I found Flossie. I say 'found', but actually I picked her up in a pub, hence her rather tarty name. It was a very crowded pub and this minuscule tabby kitten was scrabbling around people's legs and under their feet, gobbling up the odd crisp or sandwich crumb that came her way. I picked her up before she got crushed underfoot, and shared my lunch with her. She tucked in with gusto. When I offered her a sip of my sherry she lapped it up appreciatively: a cat after my own heart. As nobody seemed to know who she was or where she came from other than that she'd been hanging around the pub for some days, I took her home with me. Considering she was so tiny, she was playing host to an incredible number of free-loaders – fleas, worms, ear mites, the lot. It took several visits to the vet to get her sorted out, but the result was well worth it. She turned into a fat, contented little bundle of tabby puss. Min was delighted with her and they immediately chummed up. Rufus did his 'What-other-cat?-I-can't-see-any-other-cat' routine and went on with his serene existence.

About three months after Flossie came, Jim Barry showed me a litter of kittens just born to a stray he was caring for. Among the wriggling sausages was a pale mushroom-coloured one. Not really being in the market for another cat, I was surprised to hear myself saying, 'I'll have that one when it's big enough.' I then forgot all about it, so when Jim turned up at my flat with what looked for all the world like a tiny pink rat peering out from his pocket, I was totally unprepared. Just as all brides are supposed to be beautiful, so all kittens are expected to be pretty. But, though it grieves me to say it, this really was the plainest kitten I have ever clapped eyes on. He didn't even look like a kitten, or any sort of cat for that matter. He was incredibly skinny, with long legs out of proportion with his tiny body. His face was fox-like, with a bumpy forehead and oversized ears. There was none of that chocolate-box prettiness or wide-eyed appeal that one tends to associate with kittens. But, for all his lack of conventional good looks, there was something about Charlie's face that was absolutely captivating – it was split ear to ear in an  enormous grin! He looked just like a pixie, and his tiny pink body was one vibrating purr. He climbed out of Jim's pocket, hauled himself up to Jim's shoulder and, looking around him, purred approval of everything he saw. I put him on the floor and he immediately set off on a tour of inspection. Nothing escaped his attention, and that included the other cats. It was no good Rufus going through his 'Take-no-notice-and-it-will-go-away' act – this was one cat that was not going to be ignored. Pinning Rufus to the carpet with a paw the size of a peanut, he explored every nook and cranny, every hair and whisker. Rufus was too shocked at this outrageous liberty-taking to make even a token protest. Inspection completed, Charlie gave Rufus his seal of approval with a matey tap on the nose and passed on to the other two. Fat, easy-going Flossie took her inspection in good part, but Min wasn't about to be put upon by this minuscule upstart. She clouted him. Charlie, undaunted, clouted her back. Spitting venomously, Min shot off and took refuge under the table. Laughing his head off, Charlie trotted over to me, shinned up my trouser leg and onto my shoulder and planted a very wet kiss in my ear. It seemed I had a character on my hands.

Charlie is so full of love for everybody and everything that at times he seems fit to burst. Callers are assumed to have come expressly to visit him and are welcomed with open paws – literally. Before they even get through the door, this animated bundle of fur is climbing all over them, his purring mechanism stuck fast in overdrive. Clawing his way up to their shoulder, he grabs them in a stranglehold and rubs his knobbly little head lovingly against their faces. It never seems to occur to him that not everybody likes cats and that his attentions might conceivably be unwelcome. He is totally unsnubbable, refusing to consider the possibility that even the most committed of cat-haters might find him less than irresistible. 'You don't understand,' he tells them, tightening his hold on their necks. 'I love you. And of course you love me. Everybody loves me.' Then, completely ignoring their frantic efforts to disentangle themselves from his furry embrace, he purrs happily in their purple, oxygen-starved faces. I get quite hurt on his behalf when he is rejected, but not Charlie – he just goes back for more. The only time I have ever seen him really put out was when an obsessive cat-hater nearly had hysterics and I reluctantly had to shut Charlie out of the room. He was absolutely bewildered at what he regarded as my quite unjustified action and told me so from the other side of the door in no uncertain manner. Being part Burmese, he has a quite un-cat-like, almost human voice, and he spent the next ten minutes trying to explain that it was all a misunderstanding. All he was trying to do was to show the lady that he loved her. If I would stop being so obtuse and just let him back into the room, he was quite sure he could get it all cleared up in no time. When the coast was clear and I let him in again, he rushed around frantically, seeking the object of his affections so that he could put things right. When he realised that she had gone, he looked at me reproachfully. Now she would never know what it was like to be loved by Charlie. Remembering her near-collapse into gibbering hysteria, I thought this was probably just as well.

Shortly after Charlie's arrival at the flat, I got a phone call from Jim. Five tiny kittens had been found tied up in a sack and, judging by their condition, they had probably been incarcerated for three or four days. Two had died and another two had been found homes. The remaining one was all right physically but its nerves were shot to pieces and it would need very special care and attention if it were ever to recover from its dreadful ordeal. Did I think, in the circumstances, I could possibly see my way to . . . ? Telling myself that one cat more or less wouldn't make that much difference to my already disordered life-style, I set off to collect Cassie (or Black Pudding as he very soon came to be known). I took Charlie along for the ride, on the excuse that he was too tiny to be left alone and one of the other cats might polish him off for a snack. In truth, he loved riding in the car with me and I loved having him. He started off in my pocket, his little foxy snout peering out inquisitively. After about five mintues he decided that this position was too restricting and scrambled up onto my shoulder where he installed himself as lord of all he surveyed. Bestowing benevolent smiles on passing motorists, waving a regal paw from time to time, nodding approvingly at anything that caught his fancy, he was a real joy to be with.

When we arrived at the RSPCA Centre, I popped Charlie into his basket while I went inside. After about five minutes Pudding was produced – a tiny, completely circular, fluffy bundle of black fur. Two enormous green eyes peered out from the middle of this furry ball and he

looked just like a rather cross owl. He was quite devastating – and absolutely terrified. He cowered, trembling with fear, in the corner of his box. When I very gently put out my hand towards him, he shrank away. And who could blame him? After what he had been through, he certainly had no reason to trust anyone. 'Never mind, little puss,' I said, 'we'll make it up to you.'

I put the box in the back of the car and let Charlie out of his basket. He was absolutely agog with curiosity about the contents of the box: 'A present? For me? How exciting!' Using his pointed little snout as a lever, he gently pushed up the lid and peeked inside. He couldn't believe his eyes. A baby! His very own baby! (The 'baby', whilst admittedly very small, was nevertheless quite a bit bigger than Charlie.) Without further ado he hopped inside the box and, pinning the occupant down with one very determined paw, subjected him to a tip-to-tail licking. Having cleaned all the visible parts to his satisfaction, he prodded Pudding with his nose until he turned over, then attended to the rest of him. This done, he gave Pudding a matey pat on the snout, snuggled under his chin and promptly fell asleep. All this had taken about three minutes, I suppose, and had taken Pudding completely by surprise. But clearly Pudding was not at all put out by Charlie's attentions; burying his head in Charlie's fur, he immediately returned the compliment and treated him to a once-over-lightly with his rough little tongue. I was delighted at this quite unexpected turn of events and congratulated myself on my foresight in bringing Charlie along.

Probably because they were more or less of an age and had arrived so close together, Charlie and Pudding immediately became a unit. Where one was, there was the other. They slept together, played together, spent all their time together. Charlie, although much smaller than Pudding (who rapidly grew into an enormous great beast, hence his nickname), set himself up as Pudding's protector and mentor. I honestly think that, without

16

Charlie's support, Pudding would never have got over his early fears, because he really was the most neurotic cat I have ever come across. Everything struck terror into his heart: the phone ringing, a cock crowing, a knock on the door, even the whirring of the fan heater. Any one of these or a thousand other things would reduce him to a quivering jelly and send him scuttling under the sofa. Charlie would follow and, curling up alongside, gently comfort him. Pudding for his part idolised Charlie and this too brought its problems, because while Charlie's Incredible Courage in talking to strangers and investigating Unknown Dangers filled Pudding with awe and admiration, it also caused him considerable anxiety. What if anything happened to Charlie? He was so impulsive, always taking risks and inviting trouble. Sooner or later Something Terrible was sure to happen and he, Pudding, would lose the only friend he had ever known.

Because of his intense nervousness, I kept Pudding well away from the other three cats at first. I fed him and Charlie in one room, the other cats in another. It was when I gave Pudding his first meal that I got my biggest surprise. His hunger (or rather, as it turned out later, sheer love of food – well, greed, actually) completely overcame his nervousness and he became a raging wildcat. He demolished his meal in record time, slurping

it down noisily and with a total disregard for even the most basic elements of feline etiquette. Charlie, himself something of a dainty feeder, looked rather pained at this shocking display of piggishness but, conscious of his friend's unfortunate background, magnanimously made allowances. In between slurps, Pudding looked over his shoulder, growling menacingly. Having become used to cats who regarded anything that wasn't smoked salmon or peeled prawns as pig-swill, and who made it quite clear that they ate my inadequate offerings only as a personal favour to me, I found the gusto with which Pudding consumed his food quite heart-warming. Of course, I thought, he's been starved, so naturally he appreciates anything in the way of food; once he gets settled in he'll be just as faddy as the other resident gourmets. But I was wrong. True, he no longer growls while he's eating (he soon realised that this was a waste of good eating-time) but he still wolfs down everything he's offered as though expecting every meal to be his last. I like this as it makes life much easier for me; I put down five dishes of food and, while the other four are still tentatively sniffing at theirs, presumably checking for cyanide, Pudding has knocked his back and is asking for more. It took me a while to realise that the reason he is so undiscriminating is that he hasn't the slightest idea what he's eating – everything is gulped down so fast that he doesn't have a chance to taste it. I have never put it to the test but I'm quite sure that if I were to offer him a plate of wood-shavings and say 'Dinner, Pud', he would gobble it up and ask for seconds.

In time Pudding got over his intense nervousness and he is now a very beautiful, reasonably well-adjusted cat. But he is still very wary of strangers. If anybody comes to the cottage he is convinced that they have come to get him and he is off like a shot.

And now I was planning to uproot this motley bunch of misfits and transplant them into a strange, alien environment. Small wonder I was so apprehensive.

CHAPTER 3

As the date for the move drew nearer, so more doubts and misgivings came creeping in. Was I doing the right thing, giving up a secure, pensionable job on what I was beginning to see as no more than a whim? Swapping a comfortable, paid-for flat for a derelict hovel and a massive mortgage? Leaving all my friends to go and live amongst strangers who would undoubtedly despise me for the townie I so obviously was? What if nobody talked to me, ever? I would turn into a dotty old eccentric, holding one-sided conversations with the cats and doing an Ancient Mariner on unsuspecting passers-by. What if I fell ill? Nobody would even know – or care. Even worse, what if I dropped dead? Who would feed the cats? What if . . . ? Oh, the hell with it! Look on the bright side: no more Mrs Hoggett!

Mrs Hoggett had moved into the flat below mine some nine months earlier. She was about sixty-five and a most unlovely sight, barely five feet tall and slightly more than five feet wide. An unrestrained bosom dangled some-where below what would have been her waist if she'd had one. Her twenty-seven greasy hairs were scraped back and tied in a sparse pony-tail with a shoe-lace. She wore filthy plimsolls with bits cut out to accommodate her bunions. Teeth were worn only on very special occasions. I could have accepted all this if only she hadn't had a nature to match. Being a widow with no friends

19

(small wonder), time hung heavily on her hands, so she took up a hobby – baiting me. The slightest sound from my flat brought forth frantic thumpings on her ceiling, followed by a stream of abuse. As she never had any visitors herself, she bitterly resented anybody calling on me. Any male between the ages of eight and eighty was assumed to have come for an Evil Purpose and was greeted with screams of 'Ponce! Come to see that 'ore upstairs. Disgusting! And me a respectable widow. Filthy sods!' (My nine-year-old nephew once asked why she called me an oar. I said it was because she thought I pushed the boat out.) Women visitors fared no better. My highly respectable sister was most put out at the suggestion that she was 'another 'ore, turning the place into a brothel'.

But everybody has a redeeming feature and Mrs Hoggett had one that made up for everything else: she was the source of the most marvellous malapropisms. In the early days, when we were still speaking, she solemnly informed me that 'I am a very amicable person. I can get on with anyone. What I say is, if you can't join 'em, beat 'em. That's my Phil and Sophie.' It took me a while to realise that she was expounding her 'philosophy', and I was so delighted with the expression that I immediately dubbed my two pet toads Phil and Sophie. Since then I have always had a Phil and a Sophie among my collection of animals. Mrs Hoggett soon got the number of a neighbour who was desperately trying to hang on to her youth. 'She might dress young,' I was informed, 'but she's no spring chicken. You can tell by the crowbars round 'er eyes.' And when the people next door put up a fence of untreated wood, she portentously told them, 'That wood'll rot unless you sauerkraut it.' But my favourite was her complaint that 'I can't take all this aggravation, it gives me nostalgia pains in me 'ead.'

Surprisingly (at least to me) Mrs Hoggett was an ardent Conservative. At election time her windows were covered with posters of the Conservative candidate and pictures of her personal pin-up, Enoch Powell. Because, as I soon found out, Mrs H was a rabid racist. When a West Indian family moved in a few doors away, she nearly went berserk: 'Not enough I've got an 'ore upstairs, now I've got Darkies next door.' It was her fervent racism that gave me the idea for getting back at her. I'm not particularly proud of what I did, but I have to admit that I'm not really all that sorry.

When she heard that I was moving, she was overjoyed. 'See? God knows I'm a good woman and He's answered my prayers.'

I pointed out that she might well get someone worse.

'There's nothing worse than an 'ore.'

'Oh, I don't know,' I said. 'What if you got a black whore?'

She turned green and her toothless mouth dropped open. Leaving her to ponder on this, I went inside and laid my plans.

My West Indian friends threw themselves enthusiastically into my scheme. So it was that about three weeks later, the serenity of the smug, suburban Sunday was disrupted by an incredible scenario. Two beaten-up old cars stopped outside the gate and disgorged about a dozen West Indians arrayed in the most flamboyant attire I have ever seen outside a fancy dress gala. Two of the men were clutching guitars, one had a set of drums and another a trumpet. Some of the women were holding screaming babies, others had whining toddlers hanging onto their skirts. Maple Avenue had never seen anything like it and, to be honest, neither had I. Practically every window along the street, including mine, had an un-believing head hanging out of it as this colourful pro-cession moved noisily up the path and hammered on Mrs Hoggett's door.

If I die tomorrow I will consider my life well spent for the sheer joy of seeing Mrs H's face as she took in the tableau on her doorstep. Her mouth worked frantically as she tried to speak, but the words wouldn't come.

21

Winston, casually paring his nails with a flick knife, bared his teeth in what I think was meant to be a smile and said, 'Hi, Momma. Is this the flat that's for sale?' Treena, his wife, was peering over Mrs Hoggett's shoulder and intoning in a sing-song voice, 'Very small hall. No room for de prams. Where we put de prams?' When Carlton glanced casually up at Mrs H's ostentatious carriage-lamp and drawled, 'Yeah, a red light in this and we're open for business,' I decided the time had come to intervene before Mrs Hoggett collapsed with apoplexy. I went downstairs and said, 'Sorry you've been bothered, Mrs Hoggett. These people have come to see my flat. They must have got the wrong number.'

I hustled them into the garden, conscious of Mrs H's anguished face behind the open window, ears pricked to take in every word. Wayne was in raptures over the garden shed: 'It's perfect, man. I can strip down the wrecks in the garden and do the welding in the shed.' Gary protested loudly at this: 'Oh, no. I want that for practice sessions for the group.' So saying, he strummed a few chords on his guitar. His brother, Alvin, took this as a cue to give a discordant blast on his trumpet, while Carlton obliged with a roll on the drums. It was quite obvious that none of them had ever handled a musical instrument before, let alone played one, which was hardly surprising since Winston is a solicitor and much more at home in a Savile Row suit than in the garish, multicoloured outfit he was now wearing, while Wayne is a social worker, Gary a highly qualified teacher and Alvin an up-and-coming accountant. As for Carlton, his wicked face completely belies his gentle, caring doctor's heart. True, they were no musicians but I couldn't help feeling that they were a great loss to the stage. I took them upstairs, aware that every word and footfall was being monitored by Mrs H and would no doubt bring on her 'nostalgia'. I felt a bit mean but told myself that the end justified the means. After what she had been through that afternoon, the people who took over my flat would be welcomed with open arms and, I hoped, spared the persecution I had suffered.

As it turned out, I heard from a former neighbour that she subjected the new occupants, the Fields, to exactly the same sort of harassment that I had had to put up with – but not for long. After a week of constant haranguing, Mr Field knocked on her door and informed her, in ringing tones that carried clearly to all the eagerly wagging ears in the avenue, that one more word out of her and he would take great pleasure in parting her twenty-seven greasy hairs with a meat axe. After that there was instant peace. This gave me food for considerable thought. All my placating and reasonableness had got me nowhere with Mrs H, and it had taken a threat of aggression to resolve the problem. Much as I abhor violence, I couldn't help wondering whether world leaders and politicians were wasting their time sitting round a table and talking. Perhaps strong-arm tactics were the answer after all?

Moving day was an absolute nightmare. The Fields, having poodled around for about ten weeks before signing the contract, then wanted to move in the minute they had done so. Just before 7 am, and in a state of chaos (or, as Mrs H would have put it, 'kiosk') that had to be seen to be believed, I was astounded to find the Fields hammering on the door and demanding to be let in at once. Ignoring my pleas to let me get my things out first, they started unloading the van and moving in their goods and chattels. We kept passing on the narrow staircase, me hauling my stuff out and them lugging theirs in. Monty, a friend who was helping me move, has an artificial leg and so is at a disadvantage at the best of times when it comes to negotiating stairs. Having to manoeuvre round the Fields, who made no attempt to get out of his way, let alone offer to lend a hand, didn't exactly help his situation – or his temper. Soon words were being exchanged between him and the Fields, with Mrs H joining in the chorus from below.

I kept out of it, for I was having troubles of my own. The vet had given me tranquillisers for the cats with instructions to dose them two hours before we were due to leave, the assumption being that the actual journey would be the most traumatic part of the proceedings for them. But in fact the upheaval in the flat, the comings and goings of strangers, not to mention having their breakfast interrupted by the early arrival of the Fields, had all been too much for them. They disappeared. I was worried that I might not find them in time to give them their pills, and the thought of a long journey in the dark with five untranquillised cats struck terror into my heart. So when I came across Minnie curled up in a rug, I decided that a prematurely tranquillised cat was preferable to one that hadn't been tranquillised at all, and shoved a pill down her throat. When a second cat turned up under the bed, I repeated the procedure. The only trouble was, in all the confusion, I couldn't remember which cats had been dosed and which hadn't. But I needn't have worried, because there was no mistaking the tranquillised cats: they were the ones I kept tripping over as they lay sprawled out all over the place, snoring their heads off.

It was as well I had had the foresight to 'pack' the toads the night before, because if they'd been let loose amongst all that devastation I would never have found them, not to mention the dreadful possibility that they would have been trodden on. They had their own terrain in a corner of the room, complete with pond, mossy rocks and low-growing plants, but they liked to come out from time to time and wander around the room. As the carpet was dark green, they were virtually invisible unless they were actually moving about on it. I was so used to them coming and going as they pleased that I frequently forgot to warn visitors about them. I remember one young man who came to view the flat: we were sitting chatting over a cup of tea when suddenly, in mid-sentence, his jaw dropped, his gaze was riveted to the carpet and his eyes started to pop. Dear God, I thought, he's having a stroke. I was just about to leap on him and give him mouth-to-mouth resuscitation, this being the only thing I could think of on the spur of the moment, when his lips began to tremble and he croaked, 'It moved!' I followed his gaze and saw that Phil was emerging from under the book-case. Toads are very leisurely movers: one step, then a ten-minute pause for reflection; another step, followed by a further pause. I have known them stop in the middle of a step and stay rooted in the same position, one front leg in mid-air, for twenty minutes or more. So the poor man's reaction was quite understandable.

Someone once asked me incredulously if it was true that I actually kept toads in my flat. 'Of course,' I replied. 'Doesn't everyone?' Seriously, I have never understood why toads aren't more popular as pets, particularly for flat-dwellers who can't keep cats or dogs. They are virtually no trouble and their demands are quite modest – water to paddle in, rocks to hide under, damp moss to relax on and, of course, an unlimited supply of worms and insects. They don't smell, they don't make a mess and they really are rather enchanting little creatures. Sadly, we are brainwashed from childhood into seeing toads as repugnant, evil, ugly beasts. I blame those image-makers, Grimm and Andersen. Certainly, if any-one took it upon themselves to turn either of my toads into a handsome prince I would be most put out.

At long last everything was loaded and I was ready to go. The rain was bucketing down and it was cold and dark. The cats, having slept off the effects of their tranquillisers hours before, were bawling their heads off. They were cold, cross, nervous and hungry and they didn't care who knew it. I sympathised with them – I knew exactly how they felt.

CHAPTER 4

When the cats and I reached journey's end my spirits, which I thought had touched rock bottom, sank still further. It was pitch dark and the rain had turned to driving sleet. Because the ground was so waterlogged, it had not been possible to tow the caravan up to the cottage so it had been left at the end of the footpath, about threequarters of a mile away. This meant that there was no water, heating or light. True, there was no electricity laid on to the cottage anyway, but I had installed a small second-hand generator up there to see me through until, as the Electricity Board so quaintly put it, I could be 'energised'.

I unloaded the cats by torchlight and introduced them, with some misgiving, to their new home. They were completely bewildered. Had I gone quite mad? Had I really dragged them away from their warm, comfortable home to install them in this cold, damp, dark shed on wheels? And what was all that wet, sticky brown stuff outside? Surely they weren't expected to spend their pennies in that, soiling their immaculate paws and picking up all sorts of germs? No chance. And in fact I had to house-train them all over again, because for the first two weeks they resolutely refused to set paw outdoors, preferring to use the pot plants as a loo. The effect this had on the plants was most impressive and I seriously considered writing to ICI, telling them I had discovered a weed-killer far superior to anything they currently had on the market. When I suggested to the cats that they might consider using the litter-trays I had so thoughtfully provided, they were most affronted: litter-trays were for babies.

Food, I thought, the universal panacea. I unpacked their bowls and spooned a generous portion of minced chicken into each. Four furry faces immediately got stuck in. Rufus put on his martyred expression: 'It's supper time. You know I always have liver for supper.' I was in no mood for this sort of nonsense. My own stomach was rumbling like a volcano and for two pins I'd have snatched his chicken back and eaten it myself. I think some of what I was feeling must have communicated itself to Rufus, for he took one look at my face and decided that, just this once, he'd make the Supreme Sacrifice. As he pecked diffidently at his food, I could almost see the bubble over his head: 'Every mouthful is choking me but I'll force it down if it kills me.'

Tummies filled, the cats looked around for somewhere to sleep. There was only one place – the camp bed. My camp bed. In truth, I was grateful for the extra insulation they provided. Without bothering to undress, I lay down on the bed and wondered whether I should indulge in a good cry now or save it for later, when I could enjoy it more. I was just about to make a pact with the Devil, offering my soul (for what it was worth) in exchange for a cup of tea, when there was a tap on the door. I shot up in bed, scattering cats in every direction. Who on earth could be calling on me at this time of night? Indeed, who even knew I was here? Feverishly I shuffled all the possibilities in my mind. Burglars? Hardly – there was nothing to steal. Rapists? Also unlikely – they don't normally bother to knock. Prowlers? Nutters? Vandals? Oh, come on now – on a night like this? What the hell, I thought, whoever it is, at least it's people. I opened the door and could just make out the figure of a man standing there with the rain dripping off him, holding a torch in one hand and a box in the other. And he was smiling! I hadn't seen a smile all day.

He introduced himself as Steve Royce, saying that he lived in one of the cottages along the lane and wondered if there was anything he could do to help. Firmly controlling the impulse to throw both arms around his neck and weep on his already sodden shoulder, I invited him in. He said diffidently, 'I've brought one or two things you might find useful' and took from the box a primus stove, a lamp and – glory be – a flask of tea. Bad luck, Devil, I thought. Some other time, maybe.

Saying that if there was anything else I needed I had only to let him know, Steve left and I settled down to enjoy my cuppa. My morale had risen several notches and I mused on the magic a friendly face and a cup of tea could work. I recalled my earlier fears that no-one would talk to me; I had been here only five minutes and already I had a friend. During the eight years I lived in my suburban flat I had made only one friend among the neighbours, and that had taken five years. I decided to postpone the good cry I had promised myself. I might be glad of it later!

I lay sleepless on the camp bed all that night, listening to the rain pounding on the roof. At first light I crawled out of bed and tottered creakily to the door. The rain was still beating down and the caravan was marooned in a sea of mud. I pulled on my wellies and went out to get a better look at what I'd got myself into. In every direction, and as far as the eye could see, there was nothing but mud. I looked over towards the woods – no woods! They were shrouded in mist. 'Oh well,' I told myself not all that convincingly, 'at least you know they're there.' In truth, I didn't know anything of the kind. I was beginning to think that the whole thing was a ghastly dream. Frantically, I tried to recall all the good things about the cottage, the things that had made me buy it in the first place. The isolation? Well yes, but suddenly that seemed

28

more of a minus than a plus factor. The rolling green fields? Those green fields were now vast expanses of sloshy brown goo. The clean, fresh air? I took a deep breath of pure rainwater and nearly drowned. So instead of getting lung cancer I'd get verdigris. Big deal!

The first priority was to get the caravan moved up to the cottage. I asked around and eventually found a farmer with a four-wheel-drive tractor who, after a number of false starts, managed to haul it across the sloshy fields. I felt a lot happier once it was installed close to the cottage. If I had known then that the caravan was going to be my home for the next twenty months I might not have been quite so euphoric about it.

I had deliberately chosen March as the time to start my new life in the belief that by then the worst of winter would be over. Unfortunately, I chose the wrong year. That year winter just went on and on. Day after day there was nothing but grey skies and freezing rain. 'Think positively,' I told myself. 'Soon the rain will stop and everything will be coming up roses.' I was partly right – the rain did stop, but then the snow started. March went and April came and still it was snowing. The cats were most put out. Just as they were beginning to accept the mud as a fact of life, and indeed were quite enjoying the effect they created by bringing dirty great dobs of it into the caravan thirty or forty times a day, someone (looking balefully at me) took it into her head to replace it with this cold, white stuff. They took one look, and it was Instant Death time for the pot plants again. Wrapping myself up like an Egyptian mummy, I trudged out into the icy wastes and half-heartedly tossed a couple of undersized snowballs into the air to show the cats – not very convincingly – what fun it all was. From the caravan window five pairs of saucer-wide eyes regarded me soberly. Eventually they did venture out and were surprised to discover that it really was fun, chasing snowflakes, making snow-pies and bombarding each other with mini-snowballs.

I still had no water. I managed to find a plumber who remembered the cottage from way back and he reckoned he had a pretty good idea where the outlet was, but of course there could be no question of locating it until the snow thawed. It suddenly struck me – snow! What was snow if not frozen water, and here I was surrounded by the stuff. I grabbed a couple of buckets and started scooping up great wodges of it. When the buckets were packed solid, I just stood and looked at them. I wasn't seeing buckets of snow, I was seeing unlimited cups of tea, soup for supper, hot-water bottles – perhaps even a wash! I filled a saucepan from the bucket and got out a teabag and a cup while I waited for it to boil. That was when I got my first shock: the snow was disappearing right before my eyes. My beautiful saucepanful of snow had shrunk to an inch and a half of water. Shock number two was the discovery that even the most virginal-looking snow takes on a decidedly murky aspect once it's melted. I eyed it dubiously. Dare I take a chance on drinking it? 'Oh, don't be so prissy,' I told myself. 'It's probably a lot purer than tap water, with no chemicals or anything like that in it. Give it a good boil, fish out the odd leaf and bits of twig and you won't know the difference.' And in fact it made rather a good cup of tea – a bit earthy, perhaps, but I wasn't going to worry about trifles like that.

About three days after the snow started, I was standing by the window, watching it swirl around the fields, when I suddenly realised that I couldn't see the hedge that runs alongside the footpath. Appalled, I saw that the snow was piling up in seven-foot drifts along the path. I went out to take a closer look, and it was just like the Russian Steppes, a vast expanse of white, undulating wasteland. I was completely cut off.

Any doubts I might have had about the friendliness of country folk disappeared completely over the following weeks. Aware of my plight, neighbours rallied round and provided what help they could. Graham Christie piled his tractor with food and drove across the fields to unload it in the paddock. Alec, who lives at the farm on the other side, dragged milk and bread across on a sledge. I couldn't believe the generosity and kindness of these people, who must surely have been having problems of their own.

As day succeeded wintry day I knew, with absolute certainty, that the snow was never going to go away. The Second Ice Age had come and They were deliberately keeping it from us in case we held Them responsible and didn't vote for Them next time around. Well, They needn't worry, I thought grumpily – there's no way I can get out and vote even if I wanted to.

Then everything happened all at once: the thaw set in, the snow disappeared and suddenly it was spring. I couldn't believe it. Little tufts of green were pushing up through the ground, bare trees were prickly with buds and the birds were singing fit to bust. Even now, after all these years, I still get the same thrill when I see the first

31

signs of spring as I did that first year. When I lived in London the seasons came and went and I hardly noticed. Houses, shops, office blocks, factories – whatever the season, they always look exactly the same. Apart from wearing more clothes in winter and less in summer, the seasons made absolutely no difference to my life-style. But now that I live in the country I understand just what the cliché, 'living close to nature' means. Every day brings something new. Yesterday's bud is today's flower and even yesterday's mud is today's grass. The land-scape is continually changing, and each change brings its own special wonder. 'You see,' I lied shamelessly to the cats, 'I told you everything would be all right!'

CHAPTER 5

Life was much easier now. The water outlet had been found and I had the luxury of running – well, dribbling – water. I learned that the previous occupants had got all their drinking water from a spring at the bottom of the hill. For their other needs they had used the well in the garden, but as this was now completely bunged up with bricks, roof tiles, bits of old tree trunk and other debris, I didn't much fancy the idea.

One thing that really appalled me was the realisation that the line between civilised living and near-barbarianism is very, very thin. When I lived in London I would no more have gone to bed without a bath than I would have gone to work in my nightie. But after a few weeks without a water supply I found myself saying, 'Well, I had a wash yesterday so I can skip it today.' No wonder nobody came to visit me!

The water supply is one of the biggest problems here. The pipe bringing water from the neighbouring farm is buried a scant six inches below the ground, so every time the field is ploughed, the tractor slices through the pipe and cuts off the supply. At the best of times the water supply to the cottage is erratic. Six other households have their water piped from the farm, and if two households turn on their taps at the same time, the pressure to each is halved. If I happen to be one of the two, I get no water at all because I am the one at the top of the hill. Before I got wise to the situation and made sure I always had a couple of buckets full of water in reserve, I was forever getting caught on the hop. I still break out in a cold sweat when I remember the time, in the early days in the caravan, when I decided to update my image by putting 'highlights' in my hair. I applied the preparation according to the manufacturer's directions, then 'leave on for thirty minutes' it said. If I started heating the water on the Calor gas stove after fifteen minutes, it would be just right when the time came to wash the stuff off. But when I turned on the tap there was no water, not a dribble. I didn't panic because I still had fifteen minutes in hand. Five minutes later I still wasn't panicking; if the worst came to the worst I could rinse it off with luke-warm water. Or, as the minutes crept away, even with cold water. Twenty minutes later I decided that the time had come to panic. The longer the preparation was left on, the more dramatic the effect would be. I rushed around looking for a drop of water – anything, even a teacupful. Nothing. I eyed the cats' drinking bowl covetously but decided against it. There was no knowing when the water supply would come back and the thought of five cats screaming for a drink was more than I could stand. (Not that they ever drank from their bowl, much preferring dirty puddles or washing-up water, except when their bowl was empty of course.)

I had a sudden flash of inspiration: it didn't have to be water, it could be anything as long as it was liquid. Rummaging feverishly through the cupboard, I came across a bottle of lemonade. Perfect. No time to heat it – I sloshed it over my head, straight from the bottle. Shivering with cold, I gave my hair a brisk towelling and looked in the mirror. A total stranger looked back at me, a platinum-blonde harpy recognisable only by the look of horror in her eyes. I know I wanted a new image, but not as a middle-aged Goldilocks! The idea of exposing my blonde tresses to the public gaze appalled me, but I had to go to the shops. There was only one thing for it: I swathed my head in an all-encompassing scarf, and for the next three months my image was Mother Russia.

All this has had the effect of making me an obsessive water-miser. Not a drop of water is wasted or thrown away. Hot-water bottles are never emptied, the water is always re-cycled. Bath water is never allowed to drain away, it is collected in buckets and saved for a rainy (or rather, non-rainy) day. When I visit friends in London and see the reckless abandon with which they waste water, I just curl up. When they have finished the washing up and pull the plug out, to them it is simply the washing-up water draining away; to me, it is little short of pulling the plug on one's life-blood. No wonder word started getting around that I was becoming, well, just a little bit eccentric.

The generator was now in full working order and I had light, heating and, glory be, telly – but not all at the same time. Before I switched anything on, I had to do complicated mathematical calculations to ensure that I didn't overload the system and blow the lot. Another problem with the generator was that it needed refilling every three hours, and each night I was faced with the same dilemma. Halfway through an exciting programme, with one eye on the telly and the other on the clock, I would have to decide whether to go out and refill the generator now and miss half the programme, or to wait and take a chance on the generator running out before the end. I usually opted for the first alternative because once the generator ran out it was a devil to re-start; also, while it was still running I could at least see what I was doing, whereas if it ran out I would have to deal with it in the dark. If anyone asks me now what I remember most about those caravan days, I reply, 'Filling things.' Filling the generator, filling the Tilley lamps, filling the paraffin heater – I seemed to spend my life filling things, and I carried around with me a permanent aroma of diesel and paraffin.

There were other problems. Living in a virtual no-man's-land, I didn't expect to have my daily pinta or morning newspaper delivered. What I hadn't bargained for was not having my rubbish collected or the mail delivered. The postman was a bike-riding octogenarian and the threequarter-mile trek to my cottage was too much for him. I had to have a post-box installed at the end of the footpath and collect my mail each day. Unwelcome as bills and communications from H M Collector of Taxes are at the best of times, they are a thousand times more so when you have had to trudge nearly a mile through pouring rain to collect them. To add to my troubles, I have to fight a territorial battle every spring with the local bird population, who regard my mail-box as the ideal place to raise their young.

Then there was the mud. As the cottage is situated on top of a hill, one would reasonably expect water to flow down, away from it. Nobody believes me, but I swear I have actually seen water flowing uphill here. Obviously there must be underground springs that defy the laws of gravity. During the endless months I was in the caravan I had plenty of time to plan the decor for the cottage but, seeing the effect of twenty muddy paws on the carpets and upholstery of the caravan, I had to re-think my original ideas. Out were pastel colours, bright chintzy loose covers and gay scatter-cushions. In was any colour so long as it was dun, mud or sludge.

After several months, work still hadn't started on the cottage and I was all but drowning in a sea of red tape: planning permission, building consent, grant applications, mortgage facilities. Due to the condition of the cottage and, as they so delicately put it, because I was no longer in the first flush of youth (in other words, there was a more than even chance that I would snuff it before making the final repayment), no building society would consider me so I had to get a council loan to pay for the renovation work. It turned out that the local council hadn't even known that the cottage existed until my various applications started pouring in. An official came to inspect it and he was, not to put too fine a point on it, perturbed – not so much about the condition of the cottage as about its position. Soberly, he informed me, 'We do not like Properties to be in Isolated Positions like this. We like Properties to have a Number and to be in a Proper Street with a Name.' Crossly, he pointed out that properties such as mine were a nuisance; they were difficult to classify and this made the job of administration that much harder.

Remembering the time and effort I had put in trying to find just such a property, I can't pretend that his administrative problems caused me much heart-searching. However, he then went on to say that if the elevation had been, or became, less than ten feet above ground level, They would be empowered to demolish it and, once it was demolished, there could be no question of Their granting permission for rebuilding. This really struck terror into my heart: the cottage already looked on the verge of collapse and a gale-force wind might well finish it off, and all my dreams along with it. Being situated on top of a hill and in a very isolated position, it is particularly vulnerable to wind damage. I became obsessed with the BBC's shipping forecasts and weather predictions. Every time there was a storm, I galloped out clutching a yardstick to check that the cottage was still above the statutory ten feet. In fact, a few weeks after I moved into the caravan one of the chimneys blew down. I took the precaution of having the building shored up and slept more easily after that.

To my joy, after one or two initial hiccups, the cats settled in beautifully and took to country living as though they'd never known any other. The only one who caused me any concern was Min. She has always been something of a loner and she took to disappearing for days on end. The first time it happened, I went through agonies and spent days and nights trudging through the woods calling her. Unfortunately, Charlie insisted on

accompanying me on these forays and, as I suspected that he was probably the cause of her defection in the first place, I didn't hold out much hope of finding Min while he was around. Charlie has never forgotten the clout Minnie gave him when they first met, nor has he ever forgiven her for it. Ever since, there has been a state of war between them. Charlie bullies Min unmercifully and poor, gentle little Min, instead of standing up for herself, just runs away. She was away for over a week that first time, and I had all but given up hope of ever seeing her again when she turned up late one night, cold and hungry and bawling her head off. During the next six months she repeated her disappearing act on three or four occasions, shortening my life-span by about two years each time.

Rufus, to my utter delight and amazement, took on a new lease of life and spent his days careering around the fields like a demented teenager. Flossie turned out to be a complete dark horse. Fat, matronly, complacent Floss, whose idea of exercise hitherto was to flop off her bed, amble over to her food dish, take a leisurely mouthful or two and then collapse in a contented heap on the floor, suddenly developed an aptitude for hunting that would have done justice to a jungle cat. She has enormous patience and will sit, unmoving, for hour after hour, eyes focused on a spot where she has perhaps seen a leaf tremble or a blade of grass quiver. She knows that sooner or later whatever it is that has caused the movement will break cover, and she is quite prepared to wait until it does. Usually it is just a matter of an hour or two but I have known her wait all day for her quarry to emerge. Most times it is nothing more spectacular than a fieldmouse or a baby shrew, but she once tracked a mole along its underground route, presumably sensing its vibrations as there was certainly nothing above ground to give any clue to its presence. Once she had caught it she didn't know quite what to do with it. Moles, according to the experts, taste absolutely disgusting,

38

though how they can possibly know this unless they have actually eaten one I don't know. Anyway, Flossie certainly made no attempt to eat her catch so there might well be something in this theory. It was left to me to put the mole back into its tunnel.

Charlie, I'm pleased to say, is a bring-'em-back-alive hunter. He finds baby creatures in the fields and woods (frequently, I suspect, before they've been lost) and brings them to me for safekeeping. His first acquisition was a minute baby rabbit about the size of a bantam egg, and it seemed to me that there was no way this tiny creature would have been running around on its own. The fact that Charlie's little foxy face was swarming with rabbit fleas confirmed my suspicions that he had been down into a rabbit burrow and kidnapped the bunny. This was the first of a succession of baby rabbits he brought home, and the procedure was always the same. Having presented his find to me, he would watch anxiously while I inspected it for any injuries (there never were any). I then put it on the sofa and Charlie took over. First, a kiss on the nose to assure the baby that his intentions were honourable, followed by a no-holds-barred wash and brush-up. Nothing escaped his attentions. Every inch (and there was never much more than about four inches anyway) was licked into a stupor. Ears received special attention, inside and out. The rabbit's tail was left till last. I think it bothered him that there was so little of it – it wasn't his idea of a tail at all. The baby was then handed back to me to feed and bed down in the bathroom, this being the only room in the caravan that had a door: one cat's foster-baby could well be another cat's supper, and I wasn't taking any chances. Surprisingly, all these orphans survived – whether because of, or in spite of, Charlie's attentions I shall never know.

After a while Charlie became more ambitious and started bringing home larger creatures. If they were too big for him to carry, he chivvied them from behind, giving an encouraging nudge with his nose from time to time and miaowing his head off to give me advance warning. The high point of his career was when he brought home a ferret, although in fairness to Charlie I think this might have been an involuntary adoption on his part, because on this occasion the ferret was following Charlie instead of the other way about. 'Who's your friend?' I asked. Charlie looked embarrassed – clearly this hadn't been his idea. For my part I hadn't the slightest notion what my uninvited guest was, never having seen a ferret before. I went inside to look it up, identified it and was horrified to learn that ferrets are extremely vicious creatures who will sever a finger as soon as look at you. And I had left this bloodthirsty carnivore with Charlie! Aghast, I donned a pair of heavy gloves and dashed back into the garden. Charlie was sitting on the garden seat with his charge, putting it through his 'Everybody-who-comes-here-has-to-be-disinfected' routine. I turfed them both off the seat and sat down to watch the proceedings. Charlie promptly shinned up one trouser leg, the ferret up the other. I found a biscuit in my pocket and offered them half each. Charlie is mad about biscuits (he can hear the rustle of a packet being opened up to half a mile away) and accepted the offering with alacrity. So did the ferret. Perching himself on my knee, he gobbled it up with obvious enjoyment. I really couldn't equate this rather enchanting little creature with the bloodsucking monsters described in my book.

Of course, I realise now that he had probably been hand-reared and brought up as a pet, but whatever his background, he immediately made it clear that he now regarded this as his home. I was quite prepared to go along with this on the understanding that he lived outside. He, on the other hand, saw no reason why, if the cats were allowed inside, he should not do likewise. He soon cottoned on to the fact that the cubby-hole housing the Calor gas cylinder led into the caravan's kitchen and was in fact the cats' usual method of entry, and after that there was no way I could keep him out. My main objection to having him indoors was that he did have this, well, personal freshness problem. Not to put too fine a point on it, he ponged – hence his name, Pooh. It wasn't a grossly offensive smell, just a sort of insidious musky odour. Coming into the caravan after being outside for any length of time, I could tell immediately if Pooh had favoured us with his presence. There was also another problem: the cats, with the exception of Charlie, were very frightened of him. As soon as he set paw inside the caravan, they were off like a shot. This was fine with Pooh, particularly if his arrival happened to coincide with feeding time – which, come to think of it, it usually did. He just went from dish to dish, guzzling up the cats' food. What he couldn't eat he took away and hid, which presented another problem because his storage depot was under the fridge. I didn't fancy the idea of rotting meat and putrefying fish mouldering away there.

Eventually I devised a way of keeping Pooh out without restricting the cats' comings and goings. I kept the Calor gas cupboard closed and fixed up a stable-type door into the kitchen. The cats could jump over the bottom half, which he couldn't. I can't pretend that this was an ideal arrangement, as the kitchen door was also the outside door, so the temperature in the caravan was frequently little above freezing point. As for Pooh, I fixed him up with a nest inside the barn and he settled in happily enough. Every morning when I opened the caravan door, he was sitting on the step waiting for me. After he'd eaten his breakfast, he accompanied me on my rounds, skipping through my legs and tumbling over my feet. If I wasn't around, he latched onto Charlie, following him wherever he went. I don't think Charlie was too happy about this – even the most gregarious of cats needs some time to himself.

Pudding turned out to be an armchair hunter. Settling himself in a comfortable sunny spot, he lies down and

waits for an unsuspecting fieldmouse to walk directly under his nose. Needless to say, this almost never happens. On the very rare occasions that it does, he is so surprised that his 'quarry' invariably makes its escape long before he gets around to doing anything about it.

In fact, Flossie is the only one who ever catches anything, but even this grieves me. All the cats are well fed and have no need to hunt for food, and I hate the idea of them hunting for sport. For all that, it is possible to find excuses for them: it is instinctive, they don't know any better, they have no moral or ethical code. What I find quite impossible to justify is man hunting for sport. During my first winter here I was astounded to find the local hunt, without so much as a by-your-leave, career-ing across the paddock in pursuit of a fox. Leaving aside the questionable behaviour of riding mob-handed across somebody else's property, I happen to like foxes. Incensed, I went charging out, yelling at them to get off my land before I called the police. Of course, anyone sitting on a horse has a tremendous psychological advantage over someone standing on the ground, and these supercilious riders had an added advantage over me: they were impeccably turned out, whereas I looked as though I'd been dragged through a hedge – my working garb tends to be very basic. They looked down their highly-bred noses at me and explained that, as a mere townswoman, I couldn't possibly be expected to understand the vital role played by the hunt in protect-ing farmers' stock, particularly poultry, from foxes. I pointed out that I had cats and other assorted livestock on my land and was not too happy about the prospect of them being torn to shreds by hounds. With a contemp-tuous sneer the MFH assured me that hounds only chase moving prey. If the cats sat still, they would be quite safe.

'Oh, great,' I said. 'Perhaps you would explain to them that if they are set upon by a pack of slavering hounds, they must in no circumstances obey their natural in-stincts and make a dash for it. All they need do is stay put and all will be well.' I didn't believe it myself, so I saw no reason to suppose the cats would.

I know the fox population has to be kept down; I just can't help feeling that there must be a better, cleaner, more decorous way of doing it. The sight of grown men and women, dressed up in their ridiculous hunting gear and giving voice to infantile, ritualistic calls as they pursue their quarry, makes me quite sick. Oh yes, I know – the fox enjoys it. Being chased by a posse of huntsmen and a pack of baying hounds – what could be more fun?

Then there was the shooting fraternity. A public footpath crosses my land and I was furious to see gun-toting barbarians walking across to the woods at first light and returning in the evening bearing their ill-gotten and very dead prey. For one thing, the woods belong to the council and shooting is forbidden. For another, I was appalled to see that their bag frequently included thrushes, blackbirds, even – unbelievably – great tits, as well as pigeons and rabbits. I got in touch with the police and it was arranged that if I saw or heard anyone shooting in the woods I would telephone them immediately. After a dozen or so men had been caught and prosecuted and their very expensive guns con-fiscated, the message got through and it is years now since I've seen anyone with a gun on the footpath. A copper's nark – me? Certainly.

CHAPTER 6

Life in those early days in the caravan was full of surprises, not the least being that I had re-discovered my legs. For the past twenty years or so they had been used only to get me from the front door to the car and from the car to the office – and back again. So the discovery that they were alive and well, albeit a bit creaky, and capable of covering quite a bit of ground under their own steam came as a very pleasant surprise. Even my lungs, after years of abuse from cigarette smoke, pollution and traffic fumes, took on a new lease of life once they got used to the heady experience of inhaling clean, fresh air. These things didn't happen overnight, of course, but after a few weeks I realised that I was walking the threequarters of a mile to the post-box in one go. Not exactly a feat of Olympic proportions, I know, but previously I had had to stop twice en route to rest my aching legs and get my wind back.

As if to make up for the dreadful winter, summer that first year was wonderful: long, hot, golden days followed by warm, breeze-ruffled nights. Every waking hour I spent outdoors, just taking in the sights and sounds of the countryside. I couldn't bear to go to bed at night and end the day, and I couldn't wait to get up in the morning to start a new one. I was like a kid at Christmas, with so many goodies that I didn't know which to sample first.

Conscious of the passing of time, I was impatient of every minute wasted and resentful of anything that came between me and the world outside, such as housework and earning a living and other such boring necessities. For the first time in my life, I found myself thinking about death. Not, I'm ashamed to say, with a philosophical attitude of 'Life well spent, I'll be ready when the time comes', but with resentment and frustration. I didn't want to die – ever. It seemed so unfair, just when I'd found so much joy in life, to know that there were only so many years left in which to enjoy it. Fortunately this was only a passing phase. I soon came to realise that my frantic efforts to squeeze twenty-five hours' worth of living out of every day were getting in the way of the living itself, and after that I took each day as it came.

Every afternoon I went for a walk, accompanied by Charlie and Pudding and occasionally Flossie. Pooh the ferret insisted on being included as well, as he did in

everything. Once the other cats realised that he wasn't interested in them and that he only wanted to be with Charlie, they weren't too bothered about him. We explored all the surrounding countryside, but the woods and the Bunny Field were our favourite haunts. The woods were exciting because we never knew what we might see – a pheasant, a family of partridges, a fox or even, on one unforgettable occasion, a badger. But the Bunny Field was fun because we would sit at the top of the ridge and watch the rabbits cavorting in the valley just below. Unaware of our presence, they frisked and frolicked as though they hadn't a care in the world. The cats watched their antics with saucer-wide eyes but never made any attempt to chase them. Pooh, to my utter amazement, showed not the slightest interest in them; clearly he had never been used for rabbiting.

Someone once defined happiness as the absence of unhappiness and at one time I would have gone along with this. But I know now that happiness is a positive, almost tangible emotion. Sitting on that hilltop, listening to the birds singing and watching the rabbits playing, and knowing that I was now a part of all this, brought me a deep feeling of utter contentment. I felt at peace with the world.

So why I chose this time to introduce a totally disruptive element into my life I shall never know, but I decided the time had come to put Operation Donkey into action. In a mood of quite unjustified euphoria, I went down to Jim Barry's in Wales to collect the donkey, travelling overnight so that I could bring him back and get him settled during the day. It was love at first sight. Humphrey, as I immediately dubbed him because of his delightful (or so I thought then) tendency to rest his head on your shoulder and go 'Hummpph' in your ear, was enchanting: chocolate-brown with a creamy muzzle and liquid, black-treacle eyes. I was absolutely besotted.

The hard-core track to the cottage had not yet been laid, so Humphrey had to be decanted at the end of the footpath and led up from there. Halfway home, we were met by a welcoming committee comprising five cats in varying stages of hysteria. I had never been away overnight before and they were devastated. They thought they had been abandoned, left to fend for themselves. There would be no more loving and spoiling and walks in the woods. And, dear God, no more food! So when they spotted the Universal Provider trudging home, they went wild with excitement – until they noticed that I was not alone. They took one look at Humphrey and executed a superb collective double-take. They froze. They looked from Humphrey to me with total disbelief. Was I really planning to bring this hairy monstrosity into their nice, well-ordered lives? In fact I had been a little worried about how the cats would react to Humphrey, but figured that once he was installed in his paddock and they realised that his presence wasn't likely to affect their life-style, they wouldn't be too disturbed. I hadn't reckoned on their forestalling me. They made no attempt to hide their feelings about this latest addition to the household. They were affronted. As if it wasn't enough that I had gone off and left them, I then came back with a large, hairy beast that bore no resemblance to any cat they had ever known. It was too much. They all turned tail and stalked huffily back to the caravan, umbrage oozing from every pore.

Humphrey for his part was wounded. He had been quite prepared to be friendly and he had been spurned. Not one to take rejection lightly (as I was later to find out), he immediately took to his hooves and trotted off after them, 'hummpph'-ing furiously. Four cats shot off across the fields as though the Devil himself were after them. Charlie stopped dead and, with studied insolence, sat down in Humphrey's path. Completely ignoring the advancing juggernaut, he licked a dainty paw and passed it fastidiously over his already immaculate whiskers. Humphrey screeched to a halt within inches of a

totally unconcerned Charlie and, very tentatively, nudged him with his nose. Quick as a flash, Charlie shot out a paw and boxed Humphrey's ears. Pressing home his advantage, he then went on to tell Humphrey a few facts of life. Briefly, these were as follows: he, Charlie, was a Character and as such was acknowledged to be Boss over all other creatures, be they human or animal. This included large, shaggy things calling themselves donkeys, whatever they might be. So long as Humphrey understood this he had no doubt they would get along just fine. Bestowing a benevolent pat on Humphrey's dumbstruck muzzle, he came over and grinned at me. For my part, I was relieved to see that he had spared Humphrey the indignity of the all-over wash that he afforded most newcomers.

Meanwhile, an effectively chastened Humphrey continued quietly on his way. I opened the paddock gate and Charlie ushered Humphrey through. At first he couldn't seem to take it in but just stood by the gate and looked all around him. He took a few tentative steps and stopped. Then another few steps followed by another pause for reflection. Then he kicked up his heels and tore

off across the paddock like a mad thing. Round and round he went, head held high and tail waving in the breeze. Suddenly he skidded to a halt, collapsed in a heap and rolled over and over in a state of utter donkey bliss. I went over and scratched his little pot belly with a stick and he all but purred. Clearly there were going to be no problems here. Or so I thought!

Conscious of all that Humphrey had been through, I wanted to ensure that he felt loved and secure right from the start. To this end, I re-arranged my schedule so that I could spend the best part of that day with him. He thought this was great and, to be honest, so did I. It's not often I get the chance to poodle around doing nothing constructive without feeling pangs of guilt, so to fritter away a few hours playing silly-donkeys with an absolutely clear conscience was sheer bliss. We had a lovely time: me scratching Humphrey's ears and telling him what a super little donkey he was, and Humphrey rubbing his fat little backside against my legs in fervent agreement.

But life must go on, and came the time when I had to leave him. There was so much to do, with cats to be fed, the generator to be filled, water to be fetched. I explained all this to Humphrey and, planting a kiss on his nose, took my leave. Halfway to the caravan, I was stopped in my tracks by an ear-splitting 'WAAAGGHH!' My blood froze. I turned round and went back. Humphrey was not so much leaning on as sagging against the gate. Both ears drooped disconsolately, his lips were trembling and the overall picture was one of utter misery and dejection – for pure ham, donkeys are second to none. I fixed him with what I hoped was a no-nonsense eye. 'You called?' I enquired. One ear pricked up, one dainty hoof pawed the ground and from the very depths of his being came a heartfelt, shuddering sigh. With an air of hopelessness he rested his doom-laden head on my shoulder and snuffled grassily in my ear. 'What's the problem?' I asked. He nuzzled his head a bit deeper into my neck

49

and said 'Waaagh!' The gist of the message was that he was lonely and miserable and hurt. He had trusted me, thought I was his friend, and I had left him all alone. It was silly, he knew, to trust people. He should know better. But he couldn't help it – it was his nature. Sniff, sniff.

I hardened my heart, knowing that it would be fatal to give in to this sort of emotional blackmail. I tried to reason with him, pointing out that there were other animals needing attention. I mentioned, delicately, that if he wanted to enjoy the good life, someone (namely Muggins) had to work to provide the wherewithal. Oats, pony nuts and such-like donkey-pleasing goodies do not grow on trees. Leaving him to ponder on this, I made my way back to the caravan. I tried to shut my ears to the non-stop howling coming from the paddock, telling myself that to give in now would be playing right into his hooves. He had to learn that he could not get his own way simply by bawling his head off. When, some twenty minutes later, the sound was abruptly cut off, I said to myself, 'Aha – he's learning.' Too true: he'd learned that if the company wouldn't come to Humphrey then Humphrey must go to the company. In other words, he'd gone.

Gone? Gone where? And how? I had checked and double-checked that the fencing was donkey-proof; it was only later that I learned that a really determined donkey can get through a keyhole if he has a mind to. I got out the binoculars and surveyed the surrounding terrain, eventually fetching up on a herd of cows in the valley below. Right in the middle of the herd, and sticking out like a wart on a hog's back, was a pair of donkey ears. Off I went to collect him – the first of many, many such excursions. At first the neighbouring farmers were tolerant, even amused, but it wasn't long before they became as disenchanted with Humphrey's escapades as I was. In all fairness, I couldn't blame them. Their cows were accredited, certified, for all I know

blessed by the Pope. Infiltration by a donkey of dubious background and breeding was the last thing they needed. As for me, I soon got used to seeing a hairy, grinning face peering in at the kitchen window. For such a large, boisterous creature he could move incredibly quietly when he wanted to. I still remember the time he caught me bending in the kitchen and shoved me head-first into the oven. After that I took care to keep the kitchen door closed.

Earlier in the year I had bought thirty trees to provide a windbreak along the boundary fence. As the garden hadn't been cultivated for over twenty years, the ground was rock-hard with a dense network of nettle and thistle roots as thick as my arm. It took me several weeks of solid graft with a pickaxe to excavate the ground enough to put the trees in. Each day I tended them, watered them, spoke words of encouragement to them and greeted every inch of growth, each new leaf, with the Hallelujah Chorus. To me, those trees symbolised my new life – a settling in and putting down of roots. Then one morning I looked out of the window and saw Humphrey working his way along the row, systematically pruning each tree to about three inches above the ground. I was so choked I couldn't even speak; I just stood and looked at what used to be my beautiful trees and wept. At last I looked at Humphrey and said, 'How could you? How could you do such a thing?' He assumed his 'butter-wouldn't-melt-in-my-mouth' expression: 'Me? I didn't do anything. Not me.' As he still had a good part of the evidence sticking out of his mouth, I found his protestations singularly unconvincing.

Humphrey soon made it clear that he had two sets of standards, one for himself and one for me. It was perfectly all right, even de rigueur, for him to vacate his paddock at every possible opportunity, but I was expected to stay put. He resented it bitterly if I went out and, not being one to suffer in silence, made sure that I and everybody else within a five-mile radius knew all about it: 'She's gone,' he bawled. 'Left me all alone. Abandoned me. I'll starve to death. Call the RSPCA. Call the Donkey Protection Society. Call the . . .'

'Oh, shut up, Humphrey,' I'd say, 'I'm only going to the shops.'

Eventually I started creeping out of the caravan and crawling to the car on all fours in an effort to escape without detection. I still blush with embarrassment when I recall the time I was slithering through the mud on my hands and knees and was stopped short by a pair of blue-trousered legs about six inches in front of my face. Slowly raising my eyes, I took in a blue tunic topped by a bemused constabulary face. Thinking fast, I spluttered, 'My car keysh. I dropped my car keysh.' The policeman looked sceptical – understandably, as the keys were in my mouth and were in fact the cause of my spluttering. Resigned, I straightened up and explained the situation to him. At the time I didn't see anything particularly funny about it. Ludicrous, yes. Even pathetic in a way. But not funny. He, on the other hand, thought it was hilarious. Clutching his ample stomach, he fell about laughing. As each paroxysm gradually fizzled out, he wiped his eyes and said weakly, 'Playing hide-and-seek with a donkey! Oh, my Gawd!' and he was off again. I eyed him bleakly. To me the situation was getting less funny by the minute.

I saw him several times after that, as he used to patrol the woods with his Alsatian, 'looking for poachers,' he said. More likely he comes up for a good laugh, I thought, my paranoia getting the better of me. He always greeted me with the same words: 'Hidden from any good donkeys lately? Ha ha ha ha!' In time it got to the stage that I started hiding from him, too, if I saw him coming. That should be good for another laugh, I thought.

Humphrey also has a rather warped sense of humour. His favourite game is to creep up behind people and goose them with his nose. While they are lying face down in the mud, gibbering with shock, he gently 'hummpphs' in their ear, entreating them to get up so that he can do it again. Another favourite trick, which he never gets the chance to play twice on the same person, is to proffer his muzzle for a kiss. As soon as the victim's trusting face is within inches of his own, he very smartly brings up his rock-hard head, fetching them an eye-watering smack on the nose.

When Humphrey had been with me (on and off) for about two months, I thought it would be a nice idea to have a donkey-warming party. I hadn't done any entertaining since the move and Humphrey would provide a marvellous excuse for a bit of decorous carousing, not to mention the sheer joy of getting out of my wellies and working clobber for a few hours. I would wear a dress – I hadn't seen my legs in weeks – and make-up. Quite carried away by now, I went right over the top and decided that, yes, I would actually have a bath. This wasn't quite as simple as it sounds, not the least of the difficulties being that I had no hot water in the caravan. But I did have the Calor gas cooker, so all I had to do was boil up gallons of water on the stove, using every saucepan I possessed, and keep the water hot in the oven while I boiled up more saucepans and yet more saucepans. It's surprising how many pans it takes to provide even a couple of inches of bath water, and the first four saucepans of hot water immediately went stone-cold anyway on impact with the bath. Still, even two inches of luke-warm water, tarted up with a squirt of washing-up

liquid to give an illusion of luxury bubbles, can be a source of sheer, wanton, sybaritic pleasure when you haven't had a bath for ages.

Fortunately the day dawned bright and sunny. I wanted my friends to see the place at its best. To find themselves ankle-deep in mud, with a raw wind whipping round their delicate extremities, would only have served to confirm their firmly held and frequently expressed belief that the country was no place for people. I was delighted to see my friends again. We had a lovely time, discussing old acquaintances and places and playing the 'do you remember?' game. They thought the cottage's situation was delightful (you should have seen it a few months ago, I thought darkly) and set to with enthusiasm designing the garden or, as it was then and still is, I'm ashamed to say, the jungle. 'The swimming pool here, and a terrace there, with wide steps leading down to the lawns.' 'And an ornamental pool there, with a fountain.' 'And an orangery.' 'And a folly.' 'And a summer-house.' 'And peacocks on the lawn. Can't you just see yourself sitting on the lawn, eating wafer-thin cucumber sandwiches, with peacocks strutting all around?' Frankly no, not really. Perhaps at one time, but not now. All I could see was myself trudging around in wellies and Oxfam-rejected working clobber, up to my eyeballs in mud. They all agreed that the country was really quite pleasant, but of course they couldn't possibly bear to live there themselves.

Another reason I was glad the weather was fine was because the party was held outside. There was no way I could hold a party in the confined space of the caravan, and as work hadn't even started on the cottage, clearly it had to be the great outdoors or nothing. In any case, I wanted it to be outside so that Humphrey could join in. After all, the occasion was in his honour. He entered into the spirit of things with enthusiasm. Being a naturally gregarious beast, he was in his element in a gathering of this kind. Of course he was the centre of attention, being fussed over and fondled, having his ears scratched and his tummy tickled, all of which he accepted as no more than his due. To my surprise and relief, he behaved impeccably. He circulated among the guests, graciously accepting a canapé here, a sip of vodka there. No doubt about it, he added an extra dimension to the occasion and I couldn't help wondering why more people didn't invite donkeys to their parties.

Next morning, when I went out with his feed, I was surprised to find that he didn't come to greet me. He was lying flat out in the middle of the paddock and made no attempt to get up, not even when he saw his food-bucket. This was serious. When I went over to him I was alarmed to see that he could hardly raise his head. He opened one bloodshot eye, groaned miserably and subsided into a torpor. I was aghast. He was ill. Very ill. Probably dying. I rushed back to the caravan and rang the vet. I told him it was an emergency and he came out right away. He listened to Humphrey's chest, probed his stomach, took his temperature, looked down his mouth and into his eyes. The fact that Humphrey didn't offer even a token resistance to these indignities confirmed my fears that he really was seriously ill. Finally the vet straightened up and said, 'Well, I've examined him thoroughly and I can't find a thing wrong with him. All I can say is, he looks exactly the way I feel after a heavy night.' Oh, God, I thought. All those sips of gin and vodka yesterday!

Humphrey spent most of the day sleeping off his hangover and I took the opportunity to sign the pledge on his behalf. Never again! The next day he was fully recovered and back to his old tricks again. Time, mercifully, has drawn a veil over some of his more outrageous escapades and I no longer wake up scream-ing in the night. But something had to be done – things couldn't go on as they were. I had thought that with lots of love and attention his boisterousness would abate somewhat, but it wasn't working out like that; if anything, he was getting worse. The more secure he became, the more he was inclined to try it on, confident that he could do what he liked and get away with it. I was convinced that what he needed was a donkey companion. People were all very well, even necessary when it came to providing the good things in life, but there was nothing like another donkey when it came to companionship. Only another donkey could join you in a frolic round the field without falling down in a heap, gasping for breath. Only another donkey had a thick enough skin to take a matey nip on the backside without having hysterics. Yes, it had to be another donkey.

Now I came up against Sod's Law, which decrees that if you don't happen to be in the market for a donkey, offers pour in by every post; if you do want one, you find that donkeys have become rarer than virgins in Soho. I explored every possible donkey source, advertised in all the local and specialist papers, put the word around among friends who moved in equine circles – nothing. Meanwhile a local animal expert came up with what he was convinced was the solution: I should have Humphrey gelded. My immediate reaction was to reject the idea out of hand. Mutilating an animal to control his boisterousness (and it was just high spirits, there was no vice in Humphrey) seemed to me to be taking a sledge-hammer to crack a nut. But all the other horse-owners I spoke to agreed that gelding, quite apart from solving the behaviour problem, was in his own best interests. So did the vet when I discussed it with him, so a date was set. In the meantime I found I couldn't meet Humphrey's eye – I felt I had betrayed him. I didn't know then that the operation was going to be almost as traumatic for me as it was for him. Until then my experience of animal castration had been confined to having the cats adjusted: I simply handed the unsuspecting animal to the vet's receptionist and collected him, somewhat dozy and minus equipment, a few hours later. Nothing had prepared me for what was to come.

For one thing, I took it for granted that a general anaesthetic would be given for such a major operation but in fact Humphrey was conscious throughout. My second surprise what that I was roped in to hold his head while the unmentionable deed was perpetrated. If I'd found it difficult to look him in the eye before, it was well-nigh impossible now. The final and supreme shock came about twenty minutes later when the vet said, 'That's it, then' and shoved a handful of donkey giblets under my nose for inspection. The operation had taken place in the barn and to get Humphrey back to the paddock we had to cross the hard-core track which the workmen were in the process of laying. Halfway across, Humphrey decided he'd had enough and collapsed in a heap. With commendable presence of mind, the workmen trundled up the earth-moving equipment, scooped him up in the mechanical shovel and deposited him gently in his field.

Humphrey recovered from his operation in a couple of days. It took me a week to get over it.

CHAPTER 7

It would be nice to relate that having the Operation solved Humphrey's problem – or more accurately my problem, for as far as Humphrey was concerned, he wasn't the one with the problem. Nice, but quite untrue. Apart from a sizeable hole in my bank balance (vets don't come cheaply), I could see no difference at all. If anything Humphrey was becoming even more bolshie. Tales of his escapades, some grossly exaggerated it's only fair to say, were now so widespread that he was fast becoming a legend in the locality. I was getting quite used to complete strangers greeting me with 'Oh, you must be Humphrey's owner.' Or, more ominously, 'Oh, so *you're* Humphrey's owner.'

Not that I saw much of him these days. He spent most of his time in the next field, chatting up Joe Sheppey's cows. Plugging his escape route for the umpteenth time, I asked him crossly why he couldn't chat over the fence like most other reasonably normal folk, instead of breaking it down every time he felt a fit of mateyness coming on. He turned soft, reproachful eyes on me: It's not the same as being with them.

Well, I couldn't argue with that, but I was pretty sure Joe didn't see it that way. He was a very quiet, softly spoken little man who was too polite to say what he was undoubtedly thinking: 'Keep your bloody donkey away from my cows!' He mentioned casually that cows and donkeys don't mix; that donkeys often carry all sorts of unmentionable diseases that could be transmitted to cows; that such diseases could affect the milk yield and in time put a dairy farmer out of business. I got the message and passed it on to Humphrey. He seemed astounded to learn that he really had the power to do all that, all by himself. I realised too late that it had been a mistake telling him. Power is heady stuff – some can take it, some can't. Humphrey couldn't. It went straight to his head, and after that there was no holding him.

I decided the time had come to make peace-offering gestures to Joe, so I called on him bearing a bottle of whisky. I can see his farmhouse from my cottage, only about a mile and a half away across the fields. But to reach it by road, and in fact there is no other way, involves a six-mile trip. He invited me in and introduced me to his wife. After exchanging the customary courtesies, I diffidently proffered the bottle of whisky, mumbling something about donkey, sorry, appreciate his problem, won't happen again (ha ha) and so on. If I'd stuck a primed Mills bomb in his hand the reaction couldn't have been more electrifying. He shot to his feet, rushed to the door and, throwing it wide open, sent the bottle flying into the night. I was shattered. What had I done? I looked at Joe and couldn't believe that this was the inoffensive little man I knew – or thought I knew. He was jumping up and down in a frenzy, screaming, 'The Devil's brew! I won't have it in the house! Filthy, corrupting poison! Tempting God-fearing people into the ways of evil! What shall it profit a man, losing his soul and gaining a bottle of whisky?'

I hadn't looked at it quite like that. Clearly I had made a grave tactical error: the last thing I had wanted to do was to upset him, but there could be no doubt that I had upset him badly. Confused and embarrassed, I told him sincerely that I really was very, very sorry. Gradually his bright purple face regained its normal colour and, looking pale and drained, he sank into his chair. He explained that at one time he would have accepted my offering with pleasure. But that was before he was Touched by Jesus. Jesus had come to him while he was working in the fields one day and had put His hand on his shoulder. 'It was like a flash of blinding light and I knew my life would never be the same again.' Since then Jesus had come to him several times 'always when I was unsure and needed guidance' and shown him the right way. I am not a religious person but I was immensely moved by the obvious sincerity and depth of his belief.

As he showed me to the door, Joe earnestly entreated me to be alert and ready always for the Voice of Jesus, as he was sure He would come to me in time and change my life, too. I wasn't sure that I was up to too many changes at my age, but promised to be on the look-out. As I switched on the car headlights I noticed something gleaming by the side of the drive. I got out to investigate and saw that it was the bottle of whisky, still intact. Reflectively I picked it up and put it in the car; no doubt about it, it certainly did pay to be on the alert. On the journey home I thought about Joe and the strength of his convictions. I seriously wondered whether I should have asked him to petition God on my behalf to do something about Humphrey, but in the final analysis I knew that this was something I would have to sort out for myself.

I had listened to the experts and taken their advice and it hadn't worked. From now on I would trust my own instincts. I was still convinced that what Humphrey needed was a four-footed companion, preferably a donkey, but I was now so desperate that if anybody had come up with the offer of an elephant or a giraffe I would have accepted with alacrity and no questions asked. So when an animal sanctuary rang and asked if I'd be willing to take in a goat, I jumped at the offer. A dear little goat: why on earth hadn't I thought of it myself? The voice at the other end of the phone went on to explain that the goat had outgrown his pet status and now needed a loving home, with plenty of space and lots of companionship. By now I was so enthusiastic about the idea that I was listening selectively. I seized on the 'companionship' bit and completely missed the implications of the rest of the message. Companionship – perfect! What nobody saw fit to mention at the time was that the companionship had to be human.

Stopping only to instruct Humphrey to stay put as I had a surprise for him, I set off for the sanctuary. By now the image of Barney the goat as a gentle, mild-mannered little creature was so firmly fixed in my mind that I was totally unprepared for the reality. He was enormous! I had no idea goats came quite as large as this. Tentatively I offered him a biscuit; he swallowed it in one gulp and, by way of thanks, gave me a friendly nudge in the midriff. Gasping for air, I picked myself up off the ground and took a long, hard and somewhat apprehensive look at my latest acquisition. He really was a size, and strong with it. 'Well,' I told myself, 'at least he'll be able to hold his own with Humphrey.' I rubbed his knobbly head, which seemed to be made of solid rock. Having experienced the effects of a friendly nudge, I prayed fervently that I'd never be on the receiving end of an unfriendly one. Clutching at straws, I told myself that it could have been much worse; at least his previous owners had had the foresight (or sense of self-preservation?) to have him de-horned. I was also relieved to learn that another, more basic, adjustment to his anatomy had been made. Thank God I wouldn't have to go through that again.

Not feeling up to the journey home just yet, I thought I'd spend a bit of time getting to know Barney better. He really was a very likeable chap. Rather like a punch-drunk boxer, he was all brawn and not a lot of brain. He clearly had no idea of his own strength and couldn't understand why his friendly overtures invariably ended with the recipient lying spark out on the floor, fighting for breath. After he'd felled me for the third time and I was lying on the ground wondering whether it was worth the effort of getting up just to be knocked down again, the sanctuary superintendent came over and said, 'Ah, I see he's taken to you.' I asked him weakly how he could possibly tell. 'Well, if he didn't like you he'd be standing on your chest now. And in any case, you can see he's upset.'

I looked up at Barney. He was standing behind my head, moon-like face creased with distress, 'maa'-ing anxiously. I was absurdly touched – he *did* care!

With the goat safely loaded into the van I set off for home. Barney, having decided that he liked me, felt that he had to demonstrate his affection. Planting his front hooves on my shoulders, thereby cutting off the blood supply to my hands and arms, he browsed happily on my hair. What was going on at the other end I'd prefer not to say. I hadn't realised that goats were so prolific.

I couldn't wait to get home and introduce him to Humphrey. I had it all worked out: I wouldn't take Barney into the paddock to start with, in case Humphrey resented an interloper on his patch. I didn't want to take any chances on blighting the relationship before it had even started. Leaving Barney on the path, I let Humphrey out of the paddock and held my breath as they moved slowly towards each other. It was just like *High Noon*: slowly the gap between them narrowed and, just as I was thinking I couldn't stand the suspense a moment longer, Humphrey suddenly bared his teeth, let out a blood-curdling yell and charged. Barney drew himself up to his full height – he was bigger than Humphrey – and 'maa'-ed enquiringly. The next thing he knew, a full set of donkey choppers was implanted in his backside. I was appalled. 'Oh Humphrey,' I said, 'don't you like him?' Humphrey regarded me with wide-eyed innocence: 'Like him? Of course I like him. He's delicious!' And so saying, he took another bite.

I was bitterly disappointed at the collapse of my match-making efforts. I told myself that it was early days

yet and that once they got to know each other they would be the best of friends, but I didn't really believe it. Humphrey despised Barney and made no attempt to hide the fact that he regarded him as a complete moron. Barney wasn't all that bothered, but I was. Also, I was hurt. This was all the thanks I got for the trouble I had taken to find Humphrey a companion. 'He might at least make the effort,' I told myself pettishly. I decided to have it out with him.

As soon as Barney spotted me in the paddock he came charging over, bleating excitedly. I took evasive action but I wasn't quick enough. Lying winded on the ground, it occurred to me that I seemed to be spending a lot of time on my back these days, gazing at the sky. Turning my head, I caught Humphrey's eye and it was quite obvious that he was thoroughly enjoying all this. That did it. I hauled myself up and told him a few home truths. 'Don't you push your luck with me, you bloody-minded moke. You're a very fortunate donkey to have such a good home: plenty of food, a loving owner, a goat for company . . .'

At this last he pricked up his ears and shot Barney a look of utter contempt.

'Oh, don't be so snobby,' I snapped. 'He's supposed to be a companion, not a chess partner. Good Heavens, if all relationships were based on intelligence, some of us wouldn't have a friend in the world.'

He favoured me with a withering glance. 'You speak for yourself, mate,' it clearly said.

Barney for his part made it clear that he could take Humphrey or leave him. Most of the time he left him, literally. Too big to squeeze through gaps in the hedge like Humphrey, he simply vaulted it. The trouble with Barney was that he not only wanted to be with people, he thought he *was* people. People live in houses, so he would live in the house, or in this case the caravan. Closed doors were no deterrent – he just barged through them, using his head as a battering ram. If sledge-hammer tactics didn't work, he ate his way in. All the rubber round the door, the steps leading up to the door, finally the door itself: all went the same way.

Even more than houses, Barney loved cars. As soon as he heard a car on the path he was off at the gallop. Before the hapless driver had a chance to shut the door, he would find himself being nudged out of the way by a very determined, single-minded goat. Squeezing himself into the driver's seat, Barney would plant both hooves on the steering wheel – and I have to admit it really was a very funny sight, as long as it wasn't your car – and then look round for something to eat. All was grist to his mill: maps, Kleenex, rubber trim, seat belts, books of Green Shield stamps. Would they, I wondered, exchange him for a gift? On one occasion he gobbled down an all-but-finished report I was in the process of writing. Believing that the truth beats elaborate excuses hands down in situations of this kind, I rang the client and told him that I was very sorry but unfortunately his report would be a bit late as the draft had been eaten by a goat. A long pause. Then, 'Well, at least it's different.

Top marks for originality but none at all for credibility.'

Although he would eat virtually anything (I say 'virtually' because I once offered him one of my home-made rock cakes and he spat it out), Barney had two abiding passions – beer and cigarettes. These didn't figure too frequently in his diet as, apart from the expense, I wasn't too sure that booze and fags were any more healthy for goats than they are for humans. But as a special treat, or as a ploy to get him to do something he didn't want to do, they worked wonders.

For some time now I had been meaning to get in touch with the local police and ask them to send somebody up to advise me on security. Being in such an isolated position, it seemed to make sense. For one reason or another, though, I kept putting it off. I suppose I was afraid they would write me off as a cranky old maid, living in constant fear (or worse, hope) of being ravished. Also, they might send my old friend, the Laughing Policeman, and I couldn't face that. When eventually I got round to ringing them, they were very helpful and promised to send somebody along the following day. The track had been laid by then but it was so rough that most callers preferred to leave their cars at the bottom and walk up. It was a very hot day and when the rather portly officer arrived, puffed out and sweat-ing, he was not in the best of tempers. So far as security went, the gist of his message was that anybody who was prepared to trudge nearly a mile up a rough, pot-holed track in order to rape me deserved to get away with it. (Should I, I pondered, put up a notice at the end of the track, 'Rape by appointment only'?) I gave him a glass of beer and, as we stood chatting outside, Barney came up and gave him a matey shove. It turned out that the bobby was something of a goat-fancier and he thought Barney was rather a splendid fellow. While he was rubbing his head and telling him what a splendid fellow he was, Barney was systematically stripping every button off the front of his tunic and eating them. I just prayed that he would leave before Barney finished off the tunic and started on the trousers.

The effects of Barney's appetite were not all negative. For the past few weeks I had been plagued by an extremely militant woman who had taken it upon herself to ensure that 'PROPERTY OWNERS DO NOT TAKE LIBERTIES WITH FOOTPATHS' (she talked like that, in capitals). At least once a week my heart would sink as I spotted her plodding up the footpath, armed with an Ordnance Survey map on which she inscribed notes such as 'footpath blocked', 'farmer ploughed across footpath', 'misleading notice saying "Private Property" ' and so on. On this particular occasion she was standing by the paddock gate, map held behind her back, while she expounded on her favourite theme: 'THERE SHOULD BE A STILE HERE. THIS IS A PUBLIC FOOT-PATH AND PUBLIC FOOTPATHS MUST HAVE STILES AND IT IS THE RESPONSIBILITY OF THE PROPERTY OWNER TO PUT UP THOSE STILES.' Meanwhile, I was delighted to see that Barney had crept up behind her and was making a meal of her map.

Work had at last started on the cottage and, knowing my place and not wishing to offend the workmen, who could be remarkably touchy about anything remotely resembling interference on the part of the cretin who was paying their wages, I kept well out of their way. So I was surprised to be confronted one morning by a deputation of workmen with a Grievance. They were very sorry, they said, but they couldn't be expected to do their work properly if they were being watched all the time. I was dumbfounded. What on earth were they talking about? I never went into the cottage while they were there; instead, I sneaked in after they left each evening, pussy-footing around like a cat burglar, ter-rified that I might disarrange something that would give me away.

'Oh no,' they said. 'It's not you. It's that bloody goat. He just stands there, watching us. It gives us the creeps.

And he eats our tools, and our sandwiches, and my jacket. And Fred's cap. And . . .'

At this point I switched off. I wasn't going to tell them, but I rather liked the idea of Barney appointing himself foreman and keeping an eye on them, particularly as I was too gutless to do it myself.

Unless I was quick enough to forestall him, and usually I wasn't, Barney accompanied me on my nightly inspection of the cottage, which meant that I had to go round later with a dustpan and brush, removing the evidence. I would go upstairs and Barney would merrily trot up behind me. Upstairs inspection completed, I would then go down again, but Barney would stand rooted at the top, terrified at the prospect of negotiating the stairs he had so recklessly climbed minutes before and bleating his head off. The only way to get him down was to turn him round (no mean feat on a narrow staircase, particularly as I had to climb over him to do it) and push him down backwards. This happened every night, and still he never learned.

Like Humphrey, Barney resented it bitterly if I went out and left him. But unlike Humphrey, he didn't just tell the world about it – he came with me. Time after time I would look in the rear-view mirror and see this demented juggernaut not only trundling after me but actually gaining on me. When he reached the end of the track he would just stand there, looking lost and abandoned as he watched the car disappearing. At first the sight of his woebegone face clutched at my heart-strings, until I learned from the people living in the cottages that as soon as the car was out of sight, he turned his attention to their gardens. What he didn't eat he trampled, and if they left their doors open, he walked straight in and settled down in front of the fire. Needless to say, he was not very popular with the cottagers and neither, by association, was I. Far from solving the Humphrey problem I had acquired a new one. Couldn't I do anything right?

CHAPTER 8

By now I was quite convinced that there wasn't a donkey in the whole country who was prepared to take on the rôle of companion to Humphrey. Word had got around and all the possible candidates had gone to ground, had emigrated or were queueing up for jobs on the beach at Clacton. 'It's your image,' I told him. 'You're just going to have to learn to live with it.' He 'humpph'-ed: 'Fine PRO you turned out to be!'

Then word reached me of HAPPA (Horses & Ponies Protection Association), a society which cares for old, unwanted and ill-treated equines. Surely, I thought, they must have a donkey who would jump at the chance of a happy, caring home, even if it meant having to accept Humphrey as part of the deal? I rang them immediately and was told that, yes, they did have a donkey available, but on long-term loan only. Although they were grateful for the offer of a permanent home for their charges, their policy was never to relinquish ownership. In this way they were able to keep an eye on the animal's welfare and, if they suspected it was not being properly cared for, could reclaim it. This seemed to me to be highly commendable and I said I would be delighted to adopt a donkey on those terms.

Of course there was a problem. There always is. The donkey was in Lancashire and I was down south. The sanctuary superintendent was going to a donkey show in Coventry the following week, though, so we worked out a compromise: she would take Simon to the show with her and I would collect him from there. D (for Donkey) Day arrived and I set off with mixed feelings: joy that at long last Humphrey was to have a companion; apprehension that the whole thing would go sour. What if Simon didn't like me? What if Humphrey didn't like Simon? What if Simon turned out to be as bloody-minded as Humphrey and the pair of them ganged up on me? What if . . . ? 'Oh,' I told myself, 'do stop it!'

I had never been to a donkey show before and it was a real eye-opener. I hadn't realised that donkeys could look so classy. I mentally compared these sleek and glossy animals, with their long, almost elegant lines, with my dumpy, pot-bellied little Humphrey and decided that Humphrey was more my style of donkey. Apart from anything else, he was more my shape! But when it came to comparisons of behaviour, there was no contest. I couldn't believe that these genteel creatures were even the same species as my yobbo. 'Oh Humphrey,' I mourned, 'where did I go wrong?'

However, I wasn't there to pay homage to donkey debs, which was just as well because any resemblance between these blue-blooded aristocrats and poor, proletarian Simon was so remote as to be totally nonexistent. In fact, at first glance it was difficult to believe that Simon was a donkey at all. Not only was he the biggest donkey I've ever seen, more like a miniature cart-horse, he was also undoubtedly the hairiest. His thick coat hung almost to the ground and completely covered his face, so that it was practically impossible to tell one end from the other. If anyone had told me that this was a yak I would have accepted it without question. He just stood there, head hanging low, looking a picture of utter dejection. I walked slowly over to him, making what I hoped were agreeable donkey noises, and very carefully put out my hand towards him. To my absolute

horror, he cringed. In Heaven's name, I thought, what have they done to you? And what am I doing to you, throwing you to the wolves – or at any rate to the Humphrey, which comes to much the same thing? But I knew there was no way I was going to leave without him. I told myself that lots of love and care was all he needed, and in time we'd build up his confidence and trust again.

It was already getting late when we arrived home, and I decided to postpone the Humphrey–Simon introductions until the next day. Simon had been through enough for one day. I left the trailer alongside the caravan and let down the back, but made no attempt to coax him out. If he felt secure where he was he could stay there, but if he wanted to venture out he could do so in his own time and on his own initiative. I took him water and hay and a mash of oats and bran, bade him goodnight and with a heavy heart went to bed, where I lay awake all night. Sick with worry and full of self-reproach, I examined my motives for adopting Simon and found them wanting. All I had thought about were Humphrey's needs and, by extension, my own. If having a companion made Humphrey less boisterous, then life would be less fraught for me. I hadn't given a thought to Simon, beyond the fact of giving him a home. I hadn't considered the possible effect of exposing this shy, nervous, ill-used donkey to a high-spirited extrovert like Humphrey. God knows he'd suffered enough, without my adding to his troubles. Clutching at straws, I rationalised that this might be the best possible therapy for Simon – some of Humphrey's sense of fun and adventurousness, his sheer joie de vivre, might rub off on Simon and give him a new outlook on life. I didn't really believe it, but I was desperate to find some gleam of hope in what looked like a pretty hopeless situation.

I got up at first light and went out to check on Simon. Horror! No Simon. Shoving on my wellies, I dashed out to look for him, trying to ignore the feeling that I had been through all this before, that this was where I came in. Was I now going to spend the rest of my life rounding up two donkeys instead of one? When I got to the paddock I couldn't believe my eyes. Humphrey and Simon were nose to nose over the paddock gate, gazing into one another's eyes and mumbling silly donkey-nothings into each other's ears. I sent up a quick prayer – 'Thank you, God, but I can't stop now' – and opened the gate. Simon edged into the paddock, keeping as far away from me as possible, and he and Humphrey fell on each other's necks. Checking that the Constabulary Comedian was nowhere around, I hid behind a tree and watched them.

Humphrey was in a transport of delight. He galloped round the field, kicking up his little heels and braying his head off. Simon trotted soberly along behind him, his virtually hair-hidden eyes glowing with adoration. At this point Barney came dashing up to see what all the commotion was about. The remnants of the kitchen towel were hanging out of his mouth; I must have left the caravan door open in my panic to find Simon, and I shuddered to think of the devastation I would find when I went back. Barney vaulted the fence and, trotting over to Simon, gave him a welcoming nudge. Simon peered at him through his curtain of hair and kissed him gently on the nose. With a bellow of rage, Humphrey came charging over and flew at Barney. 'Go away,' he roared. 'Leave him alone! He's mine! She got him for me. She told me so.' Barney, too dim to take a hint, completely ignored him and gave Simon another matey shove. This was too much for Humphrey. Frantic with jealousy, he turned round and lashed out with his back legs. Very slowly the message filtered through to Barney – he was being attacked! His eyes narrowed and he reared up on his hind legs. It was a horrifying sight – he looked about ten feet tall.

Meanwhile poor Simon was looking from one to the other with bewilderment. Had he been rescued from a cruel home simply to be pitched into a madhouse? I decided this had gone on long enough and, grabbing a stick, went charging in. I had no intention of using the stick, it was simply to show the delinquents that I meant business. But sticks had obviously played a significant part in Simon's past – one look and, ears flattened, he shot off to the end of the paddock. Barney subsided immediately. Humphrey fixed me with a speculative eye. 'Don't you dare,' I warned him. 'Don't even think it.' He mumbled something about only a joke, fuss about nothing, no sense of humour some people, and stalked off in a huff. Barney just stood there looking lost and forlorn, his brow wrinkled with puzzlement, wondering why his friendly overtures always seemed to end in disaster. I took pity on him. 'Come on, Barney,' I said, 'I'll give you a ciggy.' His face brightened and he trotted back to the caravan with me. One look inside and my sympathy evaporated; the kitchen looked as though it had been hit by a hurricane. Broken crockery and glass were swimming about in the water from the overturned emergency buckets; sodden cats, having paddled around in the debris, were industriously transferring pawfuls of water all over the living-room carpet and

chairs. I scooped them up and bundled them into the bathroom – I didn't want them getting glass in their paws – and set about clearing up the kitchen. Barney's ciggy would have to wait.

The Humphrey–Simon relationship was a success from the word 'go'. They complemented one another beautifully. Humphrey needed someone he could play God to and Simon needed someone to worship, and that's what they both got. Simon simply adored Humphrey. He was his guru, his Lord and Master which, according to Humphrey, was exactly the way it should be. Wherever he went in the paddock, Simon followed. If Humphrey was out of sight for even a moment, say behind a tree or hidden in a hollow, Simon was devastated. He just stood there and howled silently. (A funny thing about Simon was that for a good year after he came, he couldn't bray. He opened his mouth and made all the usual contortions, but nothing came out.)

For about four weeks after Simon's arrival, Humphrey made no attempt to escape and I really thought I'd cracked it this time. Then I came down one morning to find Humphrey missing and Simon on the verge of collapse. He was heart-broken. After Humphrey had been gone for about an hour he resigned himself to the fact that he would never see him again. As life without Humphrey was not worth living, he decided he would die. He wouldn't eat, he wouldn't drink, he just stood gazing disconsolately at the point of Humphrey's departure. (He was a right nark, in fact. After Simon came, I never again had to wander around the paddock trying to find the three-inch gap through which Humphrey had escaped, because Simon always pinpointed it for me.) When eventually I rounded Humphrey up and brought him back, Simon went wild with delight. He cantered over to him, braying silently; nuzzling his neck, he told him how much he'd missed him and how happy he was to have him back again. Of course, all this adulation was right up Humphrey's street and he simply lapped it up.

It didn't take long for him to figure out that all he had to do to bring on a fit of the adorations was to take a powder from time to time. So he did it again. And again. And again. Each time, poor Simon ate his heart out, convinced that this time Humphrey had gone for ever. The hardest part for me was not being able to comfort Simon; he was still too wary of me to let me get anywhere near him.

The only crumb of comfort I found in all this was the fact that Simon never joined Humphrey in these escapades. He would stay behind and mourn him but he wouldn't go with him. This must mean, surely, that he felt safe and secure in his new home? Too secure to want to leave it? In those early days I clutched frantically at any sign that he was feeling more confident and happy with his lot, and there were few enough such signs. True, he loved Humphrey and enjoyed being with him, but his fear and distrust of humans was as strong as ever. When I took out their food, Humphrey would come charging over and have his head stuck into his bucket almost before I had a chance to put it down. Not Simon. He waited at a safe distance until I was out of sight before he slowly approached his bucket; clearly convinced that it was booby-trapped, he very cautiously explored it with his nose. Somewhat reassured that it hadn't blown up in his face, he took a tentative nibble, then hurriedly looked around to make sure that nobody was creeping up on him. Another nibble, followed by another security check, and so it went on. Humphrey is a no-nonsense, 'let's - get - this - lot - inside - me - as - quickly - as - possible' eater, and as soon as he'd demolished his food, he would go over to Simon, who was still going through his 'nibble-have-they-come-to-get-me?-nibble' routine, shove him out of the way and polish off his, too. It was no use my trying to intervene: one sharp word from me, even though it was addressed to Humphrey, and Simon was off like a shot. In the end I had to tie Humphrey up while Simon was eating.

At first I thought it was just nervousness on Simon's part and that once he became more confident he would stand up for himself. But in time I realised that this was the essential Simon: easy-going, uncomplaining, anything for a quiet life. 'You want my food? Have it. No hassle.' Even now, the other animals have him marked down as a soft touch. The geese pinch his mash, the cats take the bread out of his mouth, and he just stands quietly by and lets them get on with it.

At times I despaired of ever getting close to Simon. True, the no-go area he set up between us was gradually narrowing, and I could now get within three feet of him before he headed for the hills, but there was absolutely no question of actually touching him. I didn't push it, because I knew when he was ready he would make the first move and I was quite prepared to wait. But I was longing to brush those long, shaggy locks and get a good look at the face I knew was lurking somewhere behind all that fuzz.

There was still the question of Barney. He was very much the outcast. Now that Humphrey had Simon, he couldn't even be bothered to bait Barney. Poor Barney tagged along behind them, seeking any attention; even a bite on the rump was better than nothing. I think Simon would have accepted Barney into their clique but he took his lead from Humphrey. If Humphrey said 'No goat' then that was it – no goat. I tried to involve Barney as much as possible in my activities so that he wouldn't feel too lonely. Sometimes it worked, more often the result was disastrous. He accompanied us on our afternoon walks, which was fine as long as we didn't meet anybody on these excursions. I grew tired of scooping speechless ramblers out of the mud, brushing them down and explaining that he honestly didn't mean anything by it, it was just his way of being friendly. They would look at me, taking in the general déshabillé, the goat-chewed sweater, the mud-daubed trousers, the striped hair (the results of my disastrous hair-lightening experiment hadn't quite grown out yet), then their gaze would stray to my extraordinary entourage – three cats, one goat and a ferret – and comprehension would gradually dawn. Convinced I was a nutter, they would make their excuses and leave.

There was really only one answer to the Barney problem: a companion of his own. One thing I knew for sure, it would not be another goat. He was such a sociable creature, I was sure he would get on with anyone or anything. Yes, but what? I looked at his moonish face and it suddenly came to me. Of course, a sheep! But not a fully-grown sheep; if it was going to be a member of the household, then I wanted to have some hand in its upbringing and early training. In any case, after listening to the fatstock prices on the farming programmes, I realised that there was no way I could afford a sheep. But a little woolly lamb – now that was something different again. The more I thought about it, the more I liked the idea. I could hardly bear to wait till spring, when I could put Operation Lamb into action.

CHAPTER 9

One of the greatest perks of living in the country is going to auctions. Country auctions are to Sotheby's what Safeway is to Harrods, and of course this is part of their appeal. There are all sorts of country auctions but my favourite is a weekly affair in a market town about eighteen miles away, selling farm produce, eggs, plants and 'miscellanea'. This last category is an absolute joy, a vast Aladdin's cave piled with furniture, garden equipment, carpets, domestic appliances and anything else you could possibly think of, plus quite a lot you would never think of in a million years. One of my earliest purchases was a butter churn and I was thrilled to bits with it. I would now be a real countrywoman, industriously skimming the cream off my milk and miraculously transforming it into hand-moulded pats of rich, yellow dairy butter. I invested in a book of country crafts and a dairymaid-type smock (Might as well go the whole hog and look the part, I thought) and got stuck in. I immediately came up against snag number one. As I don't have the luxury of a milk delivery, I have to use this UHT long-life stuff, and there's not a lot of cream in that. But wait – what are all those things mooing in the next field if not mini-milk-factories? A word with the factory manager and I would be in business. I drove over to Joe Sheppey's, trying not to think about the last occasion I had called on him, and bought two gallons of fresh, foaming, creamy milk. I consulted my book and learned that what I really needed now was a separator, to separate the cream from the milk. Well, tough luck, I would just have to do it the hard way: 'Leave milk overnight in large, shallow basins and skim off cream next day.' Snag number two (or is it three?): I was a bit low on large, shallow basins. Frying-pans? Why not – half the trick of self-sufficiency is improvisation. But I only had three frying-pans and two gallons of milk is an awful lot of milk. Soon every surface was covered with improvised shallow basins – soup bowls, fruit dishes, tea trays, saucers, each containing a greater or lesser amount of milk according to capacity. The next problem was keeping them safe from the cats, and practically every piece of linen I possessed was called into service to shroud them: tablecloths, pillow cases, tea towels, napkins. The place looked like a mini-morgue for dead dishes.

Came the next morning and I prepared for Operation Butter. I scrubbed up as if about to embark on delicate brain surgery and then: 'Skim off the cream into a basin.' What basin? If I had a spare basin, would I have milk standing around in frying-pans, for Heaven's sake? I inspected my now diminished stock of crocks, but all I had left were cups. Putting one aside for my cuppa, I grabbed a ladle and started skimming. I finished up with one cupful of cream, which didn't seem a lot from two gallons of milk. Still, perhaps if I left the milk to stand a bit longer another layer of cream would form. I re-shrouded the containers and went back to my Milk-maid's Bible: 'Set cream aside to ripen for two or three days.' But where was I supposed to do that? There wasn't a square inch of space left anywhere, and as it was I was having to prepare my meals on the floor. What about the fridge? The 'Bible' didn't say anything about fridges, but this could well be because it was written long before such new-fangled contraptions were even thought of. I decided to take a chance and put the cup of cream in the fridge.

By the next day the idea of making my own butter had lost quite a bit of its charm. I was sick and tired of knocking over dishes of milk every time I moved. I inspected the milk for any signs of a second crop of cream, but all I could see floating on the top was a greasy-looking film, specks of dust and a very dead spider. So much for that. I skimmed the film off the top of one of the saucers and threw it down the sink. I then offered the saucer to the cats, but they turned up their noses in disdain at the very idea of drinking milk that had been hanging around for two days. I could make cream cheese from it, I thought. (I'm not proud – if it wasn't good enough for the cats, it might still be good enough for me!) But the idea of little muslin bags dripping sour milk all over the caravan was too much. I couldn't put them outside because Barney would eat them; he'd already eaten all the washing off the clothes line. In the end I emptied the containers into a bucket and tipped it out into the garden. 'It'll probably make very good fertiliser,' I told myself.

But all was not yet lost. I still had my little hoard of potential butter in the fridge. I took it out and inspected it. Golden yellow? Well, no, not exactly. More a sort of dingy grey. Probably not matured yet, I thought. Give it another day.

Next day I took this solid wodge of frozen sludge out of the fridge and left it in the kitchen to thaw. After about three hours I eyed it dubiously. It still looked a bit ashen. Perhaps the colour doesn't develop until it's churned, I thought. I scrubbed out the churn, sterilised it and then, a-quiver with excitement, poured in the cream and started cranking the handle. Nothing happened. The paddle was going round all right but it didn't seem to be making much impression on the cream. I took a closer look and my brilliant engineer's mind pinpointed the problem immediately: the paddle reached down to about four inches from the bottom of the churn, and there was only a measly inch of cream in it. The paddle wasn't even touching the surface. I realised that it would take about twenty gallons of milk to provide enough cream to transform into butter. The only way it would be a viable proposition was if I kept my own cow, and I wasn't that enthused about the idea of home-made butter. The cats graciously accepted my offer of the cream and I sat down and did my homework. According to my calculations, if I had succeeded in producing half a pound of butter it would have cost me about £3.50. You can relax, United Dairies, your market share is safe.

But once I got the hang of auctions, I did make some pretty good buys and in fact the cottage is now entirely furnished and carpeted with auction goods. ('And looks like it, too,' one of my snooty London friends commented.) Before I learned the rules of the game, however, I ended up with some disastrous 'bargains'. People are always bringing me injured birds and animals they've found in the woods, and providing cages for them is something of a problem. Old cupboards and sideboards go for about 25p at auction, and I started buying up all I could lay my hands on. By taking out the door panels and replacing them with wire netting, these pieces of furniture could be converted into rather splendid cages. Soon, every time an old sideboard came up, the auctioneer would look to me and, before I knew it, it was mine. Came the time when I had enough potential cages to house the entire population of London Zoo and still the auctioneer went on knocking them down to me. Experts are forever warning amateur auction-goers not to catch the auctioneer's eye by mistake, but what they don't warn you about is the danger of letting the auctioneer catch your eye. Every time an old cupboard came up I would feel the auctioneer's eye fixed on me and, however much I willed myself not to look at him, I found myself hypnotically drawn to his gaze – and I was the not-so-proud owner of yet another ramshackle old piece of plywood. After a short time the auctioneer realised that he had a real soft touch in me and started looking in my direction every time he had a bit of old tat that he couldn't get a bid for. It was quite ridiculous, and most of it I left there because it just wasn't worth taking away. I once asked Martin, the auctioneer, what he did with all his old junk when I wasn't there. 'Oh,' he said, 'no problem. We just put it by until the next time you come!'

Martin also did the plant auction and this was hilarious because he knew absolutely nothing about plants. So it frequently happened that a disappointed buyer would find that he had bought a box of antirrhinums when he was under the impression that he was bidding for anthuriums. Or violas instead of violets. When he was really stuck, Martin would look to the congregation for enlightenment. If Jack was there he was all right, because Jack was a nurseryman and had forgotten more about horticulture than Martin will ever know. But if Jack wasn't there the Wing Commander assumed the responsibility of 'setting you right, old boy'. The Wing Commander was a bit of a joke, a little bespectacled weedy chap whose jolly-whizz accent slipped from time to time, revealing the thick Brum just below the surface. Whatever the weather, he wore an RAF greatcoat several sizes too large, courtesy of Oxfam, and a flying officer's scarf. Even his tatty old van was decorated with RAF symbols. As one of the auction regulars put it, 'If he's a wing commander, I'm the Messiah.'

The fact that he knew even less about plants than Martin did in no way deterred the Wing Commander from spreading the horticultural gospel. Martin would hold aloft, say, a salvia and announce, 'Now what we have here, ladies and gentlemen, is a . . .' (consulting his sales sheet) '. . . well, I can't quite make out what we have here. But it's a plant, a very nice plant, and there's twenty-four of them in the lot.' Then hurriedly, before the WC could put in his spoke, he would go on, 'Now, what am I bid for this splendid . . .'

But it was no good. Before he could finish, the WC would pipe up, 'Now, just a minute, old chap, I think I can help you out there.' Groans all round. 'When I was overseas, my old chum the Ambassador had these in the grounds of his residence. In fact he asked my advice about them and I was able to inform him that they are called Coleus. They are hardy perennials with yellow blooms.'

Even Martin knew that this was absolute rubbish and, looking dubious, said, 'Well, I don't know. As far as I can make out from the sheet, it begins with an "S".'

'Yes of course, old man, that's the Latin name.'

At this point some of the hardier perennials among the buyers could stand it no longer. 'Get back in your Spitfire and Focke-Wulf off' was just one of the less offensive suggestions made to him. Bristling with indignation, the Wing Commander would mutter something about 'no use trying to educate these ignorant peasants' and withdraw. It was all part of the fun and we genuinely missed the Wing Commander if he didn't turn up.

Even more fun than these country auctions are farm auctions. They take place all year round, but the open season really starts in September, after the harvest has been gathered in. The pressure is off farmers for a while and the atmosphere is very much like the end of term. Farm auctions fall into two categories, livestock and deadstock. After just one visit to a livestock auction, I learned to avoid them like the plague. The memory of those cows and sheep, huddled together for comfort, haunted me for days and I found myself lying awake at night, worrying about their ultimate fate. Deadstock auctions are something different again. 'Deadstock', I was relieved to learn, is not dead animals but farm machinery and equipment. Of course I was not in the market for tractors or silos, but there are also all sorts of other handy items such as buckets, boxes of nails, fence-posts, farm gates and the like. Since the advent of Humphrey and Barney I have become very security-conscious, so fencing materials are always of interest. All the lots are laid out in a field and the auction takes place on site, so it helps if the weather is fine.

The start of the auction is heralded by a little man in a white coat ringing a handbell. This is the signal for all the poverty-stricken farmers who are swapping hard-luck stories and bemoaning the disastrous harvest in the bar (farm auctions always have a licensed bar, which adds to their appeal) to swallow their double brandies, collect their suède and sheepskin coats from their brand-new Range Rovers (I was mortified to note that at these farm auctions, my vehicle was always the oldest and tattiest) and make their way over to Lot 1. I love Lot 1. It usually comprises all the old rubbish which defies individual classification and so has been lumped together as one lot. Separately, none of these items would rate a second glance but piled up in a heap, they take on the appeal of treasure trove. Lot 1 is usually bought by a scrap dealer and is invariably followed by a second, entirely unofficial sale, with people sidling up to the purchaser and whispering, 'Pssst! Want to sell that kettle/bit of barbed wire/chicken feeder?' It often turns out that once these private transactions have been completed, the original purchaser has covered the cost of the entire lot. With Lot 1 out of the way, the auctioneer then gets down to the serious business of the sale.

There were lessons to be learned here, too, the most important being never ever to bid for anything that I couldn't get into the back of the van, no matter how cheap it was. In fact, the bigger the item, the cheaper it tended to be, because transport charges are so high that they completely wipe out any saving made on the purchase price. I have had to leave more than one 'bargain' behind because I couldn't get it home. I also learned that auctioneers are not above playing the occasional dirty trick and sneaking in the odd small animal among the 'deadstock', which is how Henrietta became part of the household.

Henrietta was a bird – a very large bird indeed. Speculation about her breeding and possible forebears was rife and insulting to the point of slander. The general consensus was that she was probably a cross between a peacock and a guinea fowl. Certainly she was no beauty: her little, conker-shaped head with its rheumy eyes perched above a scraggy neck was much too small for her body. She had tweedy grey feathers with splodges of white, and all in all she looked as though she'd been put together by a committee. She really was pitiful, cooped up in a makeshift cage with no room to spread her wings, and clearly very frightened by the laughing, jeering crowds around her. I grieved for her, but the idea of buying her never crossed my mind. As the auctioneer was coming up to her number (and I now appreciate the full significance of the expression, 'your number's up'), I couldn't help overhearing some of the comments of the people nearby. 'Make good eating, that one.' 'Aye, knocking on a bit, but plenty of meat on her.' I was stunned. They were going to eat this funny old lady! I don't remember much of what happened after that; all I know is that five minutes later Henrietta was mine for the princely sum of two pounds.

When I inspected my purchase more closely, my heart sank. She looked about 300 years old and I didn't see how she could possibly survive the journey home. I didn't know then what a game old bird she was. It wasn't until I got her home that it dawned on me that I had nowhere to put her. I hadn't planned to buy a turkey-sized bird, or any other bird for that matter, so I hadn't organised any accommodation. Not that I intended to keep her permanently shut up – the whole point in buying her was to ensure that her declining years (or more likely weeks, by the look of her) were as happy as possible, and for any creature that had to mean freedom. But she had to be kept in for a few days until she became acclimatised, otherwise there was a risk of her flying off and falling foul of some gun-happy maniac. And even when she was sufficiently settled to run free during the day, she would still need a secure, fox-proof place to sleep in at night. The first priority, though, was to get her out of her ridiculously inadequate cage and give her a chance to spread her wings.

I took the cage into the barn, opened the top and put water and food nearby. She would be all right there for a bit while I converted one of my auction sideboards into a makeshift home. When I went into the barn about ten minutes later to get some staples, she was sitting majestically on top of the cage, preening her feathers. She looked up incuriously when I went in, then continued her toilette. I went out and was about to shut the barn door when she swept regally past me (for all her tattiness there was a certain stateliness about her, a sort of impoverished gentility that suggested she had known better times) and out into the garden. O Lord, I thought, I'm going to lose her before I've had her five minutes. She continued her stately progress across the garden until she reached a large, felled tree. Metaphorically hitching her skirts, she hopped daintily on to it, fluffed out her feathers and settled down to survey her surroundings. 'Well, good for you,' I said. She acknowledged this with a gracious nod of her head.

There she sat for the rest of the day. After about ten minutes she was joined by Charlie, who presumably felt it his duty to make this newcomer feel at home. What Henrietta felt about this I couldn't say, for her inscrutable face gave nothing away. Charlie eyed her speculatively. Surely, I thought, he wasn't planning to give her the customary wash and brush-up he accords to all new arrivals? But after taking in their relative sizes, Charlie, who is nobody's fool, obviously decided that this was one occasion when discretion was decidedly the better part of valour. He settled for a matey kiss on the beak, which Henrietta took in surprisingly good part, and settled down companionably next to her. I was delighted at the way things were working out. Clearly Henrietta

wasn't going to need an acclimatisation period; probably Charlie had told her that, if she played her cards
right, she was on to a good thing here and she realised
that as a home, this was as good as any.

But there was still the question of sleeping quarters,
and it would be dark soon. I re-applied myself to the job
of transforming the sideboard into a temporary bed-
room, trying to ignore the thought that was niggling
away at the back of my mind: how in Heaven's name was
I going to get Henrietta into it? After being cooped up all
day and then enjoying a brief period of freedom, it didn't
seem likely that she would view the prospect of being
shut up in a converted sideboard with any enthusiasm.
As it turned out, the question didn't arise – she had
already decided where she was going to sleep. As soon
as the light started to fail, she descended gracefully from
her log and walked over to the apple tree, a gigantic tree
which is well over a hundred years old. Speculatively she
walked all round it, stopping from time to time to gauge
the distance to the lowest branch. Head on one side, she

made her calculations: height of tree × weight of bird + wing-span − negative factors such as failing sight, creaky leg-joints, lack of resilience etc. After a number of false starts she eventually heaved her considerable bulk into the air and pancake-landed on the lowest branch. From there it was a simple matter to hop from branch to branch until she reached the one she fancied. Tucking her head under her wing, she settled down for the night.

From that first day, she never made any attempt to wander beyond the bounds of the garden and I always knew where to find her at any time of the day. She was very set in her ways, hating anything that disturbed her routine. There was something rather prim and old-maidish about her that I found quite disarming. After two or three weeks she started filling out and her neck lost a lot of its scragginess. Her eyes were much brighter and her feathers took on a silky sheen. She would never be a pretty bird, for her shape was too ludicrous, but she developed a sort of splendour that to my mind set her above mere good looks.

Having Henrietta brought an unexpected bonus. Every morning I would open the kitchen door and call, 'Breakfast, Henrietta!' She came swooping down from her tree – and so did every other bird in the vicinity. It took them no more than a day or two to realise that breakfast for Henrietta meant breakfast for them, too. My morning routine is always the same. I get up, light a cigarette and start coughing. Very soon Henrietta and all her hangers-on got the message: if Mother Bountiful is coughing, she's up. After that, they didn't wait for the 'Breakfast, Henrietta' signal. At the first cough they were there, waiting.

I knocked up some rough and ready bird-tables and put out food for the wild birds. Every tree was festooned with lumps of fat, strings of nuts, half-coconuts and various other delicacies. I made little puddings for them from melted fat, raisins, pieces of cheese and cake, chopped nuts and brown bread. The cats and I spent

hours watching the comings and goings of these feathered free-loaders — chaffinches, thrushes, robins, tits, greenfinches, yellowhammers, buntings. When winter came and the weather became colder, pheasants, partridges and moorhens would come seeking sustenance. I don't know how word got around; I suppose there must be some sort of avian grapevine. Watching a pair of kestrels hovering over the paddock one wintry day, it occurred to me that birds of prey must have a hard time finding food when the weather is really bad. I set up a separate feeding area for them, putting out meat and liver and tinned cat and dog food, and was delighted when kestrels and sparrowhawks came swooping down to feed. I hoped that in the summer the birds would show their appreciation by leaving my fruit trees alone. Some hope!

CHAPTER 10

One of my favourite stories is about the New York salesman who was driving to another town and somehow took a wrong turning. He found himself driving through mile after mile of open countryside with not so much as a house or shop or telephone in sight. Eventually he came upon a shack at the side of the road; sitting in a rocker on the front porch was an old man smoking a pipe. After getting directions, the salesman asked the man, 'How come you're living in this godforsaken place?' The old man took the pipe out of his mouth, looked all around him at the rolling hills, the lush fields and majestic trees and replied, 'Just lucky, I guess.'

There isn't a day that I don't look at my own little bit of Heaven and say, 'Just lucky, I guess.' Cynical friends say, 'Oh, you're still in the honeymoon period. You'll soon get fed up with it, once the novelty wears off.' But I find that as time goes on and I get more and more involved in country life, so the pleasure intensifies.

As that first glorious summer reached its peak, I was overwhelmed by Nature's sheer prodigiousness. The hedgerows were bursting with fruit – blackberries, elderberries, rose hips, crab apples. It was a wonderful sight, but it bothered me a little because there was so much waste. Even allowing for the birds eating their fill and the smaller creatures laying in stocks for the winter,

89

the greater part of this bounty was going to be left to rot. There must be something I could do with all this fruit, I thought, apart from making interminable blackberry and apple pies. Bottle it? Make jam? I couldn't work up much enthusiasm over either idea. But why not wine? Excitedly I leafed through my country crafts book and found a whole section on wine-making. Remembering the butter-making fiasco, I determined that this time I would make sure I had all the necessary equipment before embarking on the project. As far as I could make out, the main requirement was buckets – lots and lots of buckets. During the next two or three weeks I relinquished my auction status of Chief Sideboard-Buyer and became undisputed Queen of the Buckets. When I had amassed sufficient buckets to make enough wine to float a battleship, I judged I was ready.

I decided to start off with elderberry wine, which had a nice rustic ring about it. I consulted my book: 'Gather five pounds of elderberries.' No problem there, the garden was full of them. I started gathering and gathering. And gathering. Elderberries are a bit on the small side as berries go, and five pounds represents an awful lot of elderberries. They are also inclined to part company from the bush complete with stalk, which means that every berry has to be gone over and the stalk removed. It took me nearly a day to collect the berries and another day to de-stalk them, and I was still only on line one of the recipe! I get a bit irritated with these writers who airily dismiss two days of solid graft in half a dozen words. However, I now had my elderberries and the next step was to wash them. I turned on the tap. Surprise, surprise – no water. Aha, I thought, bringing in my emergency buckets, you don't catch me like that, not a second time. I washed the berries thoroughly, carefully fishing out the dead and semi-conscious insects that floated to the top and giving them to Henrietta. Next step: 'Cut up one pound of raisins and add to the elderberries.' I have heard stories of delinquent soldiers

being put to work scrubbing walls with toothbrushes and cutting grass with nail-scissors; what I can't understand is how the sadists who think up these soul-destroying chores could possibly have overlooked the most exquisite form of torture of all – cutting up raisins. Because raisins are virtually uncuttable. They have a life of their own and, as soon as the knife touches them, they jump out of the way. If you are quick enough to catch one and actually sever it, both halves immediately attach themselves firmly to the knife and have to be levered off. Incredible, I thought, hacking madly away: you pay vast sums of money for tubes of so-called glue and it hasn't half the sticking power of a humble raisin. It took me half a day to mangle those raisins (you couldn't call it cutting) and my hand was a mass of blisters. Then 'Boil up a gallon of water and pour over the fruit.' A gallon of water? On my tiny stove? It took half an hour to boil a pint. I started boiling. And boiling. And boiling.

The caravan took on the atmosphere of a Turkish bath. Weak from loss of water and body salts, I stripped to the waist and tried not to think about my vastly depleted emergency water supply. Once the boiling water was poured on, the next stage was comparatively easy – 'Cover the bucket and leave it in a warm place for fourteen days' – except that there are no specifically warm places in a caravan. If the weather is fine, the whole place is like a sauna; if it's cool, it's like living in an igloo. There is no happy medium. Also, space is at a premium, and a two-gallon bucket that can be tucked away in any one of a dozen places in a house is as intrusive as a cement-mixer in the confined space of a caravan. My shins were covered in bruises because every time I moved, I fell over that blasted bucket.

After two weeks the liquid had to be strained off and, oh dear, 'Bring to the boil'. This time it was even worse – a fruity Turkish bath! The atmosphere was so heady that I got tipsy just breathing. When it came to sterilising fermentation jars, funnels, air-locks, spoons and count-less other pieces of equipment, I was reluctantly forced to the conclusion that a caravan with an extremely erratic water supply was perhaps not the ideal set-up for wine-making.

There was a happy ending, though, because the wine turned out really rather well and once I moved into the cottage I went in for wine-making in a big way. From blackberry, elderberry and haw (shades of Mrs Hoggett!), I went on to potato, rice, carrot, parsnip, rhubarb and marrow wine. But my favourite is tea wine. It's very easy to make, the main ingredient is readily available and it really is absolutely delicious. I have given the recipe at the end of this chapter for anybody who wants to have a go.

Whenever the subject of home-made wine comes up, I always think of Nigel, a friend from my London days, although I know he would be horrified to think that he could possibly be associated with anything quite so bourgeois and homespun as home-made wine. He is a rather precious and, I have to admit, somewhat pretentious poseur who sees himself as a sort of latter-day Noel Coward. He is the only person I know who goes to Harrods specifically to buy potatoes. The first time he visited my London flat, he stood in the middle of the sitting room, looked all around him and drawled, 'Well yes, sweetie, you do have considerable taste.' Simpering self-effacingly, I was just about to say something suitably fatuous when he added, 'What a pity it's all so bad!' He is definitely not a caravan person so I waited until I was installed in the cottage before inviting him out. I can't pretend that he accepted my invitation with unbridled enthusiasm. Of course, he would love to see me again, ducky, but – the Country? Was it, he asked himself, really his thing? So crude and earthy. The mud would ruin his Gucci shoes. And the peasants – salt of the earth, of course, but really . . . I told him that the cottage had once been part of the estate of our friendly neighbourhood duke and that one of my neighbours came from a titled family. Being an unmitigated snob, he couldn't resist this and graciously condescended to accept my invitation. I was delighted because, for all his airs and graces, he is great fun and I am genuinely fond of him.

Clearly he had given a great deal of thought to the question of What the Well-Dressed Man-About-Town Wears in the Country to Impress the Natives. I don't know about the natives, but speaking for myself I was most impressed by his deerstalker cap, tweed cape, riding boots and breeches. I couldn't believe that he had gone to the expense of buying this outfit just to visit me; he must have hired it. I wondered what had happened to the springer spaniel that must surely have been part of the deal. He thought the cottage was 'Charming, sweetie. So absolutely you.' Looking around at the hodge-podge of auction-acquired furniture and the pussy-paw-marked upholstery, I wondered if he was trying to tell me something!

With hindsight, I realise it was idiotic to offer him a glass of home-made wine. He raised an exquisite eyebrow and murmured, 'My dear child, I am too devoted to château-bottled vintage wine to ruin my palate with home-brewed fruit juice.'

By now I was getting a bit fed up with him and, nettled, I snapped, 'Well, I like caviar and lobster but it doesn't stop me enjoying the odd kipper.'

He conceded that I had a point and agreed to sample a glass of elderberry. He held the glass to the light and said, 'Hmm, quite a decent colour.' He then sniffed it delicately for a moment or two and pronounced it 'lacking in bouquet'. Finally he took a sip, rolled it around in his mouth and looked into the middle distance. After what seemed an eternity, he looked speculatively at me and said, 'You say you gathered the elderberries from your own garden?'

'Yes,' I replied.

A reflective pause. 'And you made the wine in the kitchen?'

I nodded.

A long pause, then: 'Hmm, doesn't travel, does it?'

Recipe for Tea Wine
1 gallon tea (without milk!)
4 lbs granulated sugar
1 lb raisins
4 lemons
2 oranges
1 teaspoon wine yeast
1 teaspoon yeast nutrient

Make up one gallon of tea in a bucket (I use about 30 tea bags). Slice the lemons and oranges thinly, cut up the raisins (I cheat – I mince them), and put them all in a plastic bucket together with the granulated sugar. Strain the tea onto them and mix well to dissolve the sugar. Leave it until luke-warm, then add the yeast and yeast nutrient and stir again. Cover the bucket with a clean cloth and leave it somewhere warm for about a month. Then carefully remove the scum and strain the liquid into a fermentation jar. (Squeeze out all the juice from the fruit and strain that, too.) Fit a fermentation lock and leave it somewhere warm until it clears and fermentation has finished.

Don't forget to sterilise everything first.

If you're like me and hate waste, you can use the raisins for baking – they add a nice boozy flavour. I made the mistake of feeding some to the chickens and ducks, and for the rest of the day they were reeling about all over the place like drunken sailors!

CHAPTER 11

It was late summer and I was working in the caravan, fretful at having to be indoors when the sun was shining and trying not to remember the good old pre-Barney days when I could have taken my work outside, when suddenly everything went black. Well I never, I thought, a total eclipse of the sun. I looked out of the window and saw a heavy pall of thick black smoke hanging over the garden and paddock, completely obliterating the sun. Fire! The field on the other side of the footpath was ablaze – it was stubble-burning time. I rounded up the cats and shut them in the caravan, hustled Henrietta into the barn and carried a vociferously protesting Pooh into the shed. When I went back outside to calm Barney and the donkeys, who were rushing around in a frenzy, I noticed that part of the hedge was alight. Just then a fire engine came rattling up the path. The men jumped out and unrolled their hoses. 'Quick!' they yelled. 'Where's the water outlet?'

I pointed to it and started, 'But . . .'

'No time for chat,' they said. 'Got to get this under control before the whole hedge goes.' They plugged in, turned on the tap and aimed the nozzle. A pathetic dribble of water trickled out and dripped slowly onto their nice shiny boots. Speechless, they turned to me.

'I tried to tell you,' I said. 'The water pressure here is

virtually nil.' As they trundled across to the farm, I realised for the first time that if the cottage or caravan ever caught fire I would be in real trouble. I went inside and added 'fire extinguishers' to my shopping list.

Stubble-burning has become a hotly contested issue in the county over the past few years, causing a rift between farmers on the one side and local people and conservationists on the other, with me firmly entrenched somewhere in the middle. I can fully understand and sympathise with the farmers' point of view that burning is the most efficient and effective way of getting rid of the stubble and at the same time sterilising the soil. Also, his fields are in effect the farmer's workshop, and how he carries on his business is surely his own affair. Nobody would dream of telling, say, a shoe-maker or toy manufacturer how to run his factory. At the same time, we do tend to feel that the countryside belongs to us all, that the farmer only holds it in trust and that anything that blights the countryside threatens our heritage. There is also the question of wildlife: a burning field brings death to countless numbers of small creatures and renders many more homeless. I honestly don't know what the answer is and, as a countrywoman by adoption only, I really don't feel I have the right to lay down the law on issues affecting people who get their living from the land.

Farmers have become the target for a great deal of criticism in recent years, mostly from people who know little and care less about the problems of farming. The so-called 'impoverished farmer' is a standing joke amongst townspeople, and it may have become obvious that I'm as guilty as the next person, but whether the average farmer is as well-heeled as legend suggests or not, I do know that he earns every penny he gets. During the season my farmer neighbours are out working in the fields at first light, often carrying on right through the night. And of course it is a very precarious business – a late frost or wet summer and all their work has gone for

nothing. My respect for farmers has increased enormously since I came to live amongst them.

I am talking now about real farmers, not the city gent
who buys a country property with a few acres, usually
for his daughter's pony, and sets himself up as a
'gentleman farmer'. He usually turns out to be neither.
Although he is generally loathed and despised by
genuine farmers, he does serve a useful function – he fills
the gap left by the now defunct village idiot.

You can recognise a pseudo-farmer on sight because
he always Looks The Part – or so he thinks. Actually, no
genuine farmer would be seen dead in the check cap,
hairy ginger tweeds and shiny boots that the pseudo
affects. Suitably attired, the next step is to Act The Part.
This entails patrolling the bounds of his Property (to the
pseudo-farmer, his Property always has a capital 'P')
morning and evening. Essential accoutrements are a
knobbly stick, a gun and a sporty-looking dog. An
important ritual to be observed during these walks is for
the pseudo to say 'Heel, boy!' at regular intervals and for
the dog to ignore him. The delights of the walk are
considerably enhanced if the pseudo meets a Genuine
Peasant during the course of it. This doesn't happen
often because Genuine Peasants run a mile if they see
him coming. If he is lucky enough to catch one, he
immediately assumes the rôle of squire (a neighbour
once described our local pseudo as suffering from an
acute case of squiarrhoea), addressing him as 'My good
man' and 'My dear fellow', and is most offended when
the GP doesn't even have the grace to tug his forelock. He
goes home and complains to his wife that it's no use
trying to be friendly to these yokels, the only thing they
understand is a Firm Hand. Mrs Pseudo listens with only
half an ear because she is having troubles of her own: she
has made it clear to all the local organisations that she is
quite prepared to sit on their committees and give them
the benefit of her superior knowledge, on the understanding that she is made President of course, and

cannot fathom why they are not falling over themselves to take advantage of her generous offer.

After brooding for a while, the pseudo decides to Make His Presence Felt. He puts fences round everything. (There is a story locally, undoubtedly apocryphal, about the wife of an obsessive pseudo who wasn't quick enough off the mark and found herself fenced in, together with her two children and her mum who happened to be visiting.) The fencing-off is followed by a rash of notices proclaiming 'Private Property', 'Keep Out', and 'Trespassers Will Be Shot'. This last notice does have some effect. The locals all know that he wouldn't have the nerve to shoot at them – he'd fire to miss. But as it's painfully obvious that the pseudo has never handled a gun in his life before, the chances are that he would hit them, and whether you're hit deliberately or accidentally, you tend to end up just as dead!

The fact that he knows absolutely nothing about farming in no way deters the pseudo-farmer from giving the genuine farmers the benefit of his ignorance. He is always ready to offer them gratuitous advice about the best way to farm, the best seed to sow, the best fertiliser to use, and is genuinely perplexed when their crops turn out to be far superior to his own sickly, virus-infected efforts. What he doesn't know is that his land is virtually useless for farming – if it was any good the local farmers would have snapped it up.

Even worse are the townspeople who move to the country and straightaway start imposing their urban standards on the rural way of life, insisting on street lighting, objecting to farmyard smells, complaining about crowing cockerels, demanding all the mod cons of town life such as theatres and launderettes but strongly opposing any attempt to site these 'necessities' within shouting distance of their own homes. I honestly can't see the point of escaping from suburbia if you are then going to change your new environment into a cheap replica of the one you left behind. I was talking to a farmer at an auction a few weeks ago and he told me that a London couple had recently bought a cottage overlooking his farm. After, as he put it, 'tarting up the place like an olde-worlde bingo hall', they called on him to complain about his newly ploughed fields. All that bare, brown earth looked so depressing, they said, and completely ruined their view. Would he mind planting something green in it, please?

CHAPTER 12

There was a definite nip in the air these mornings and I realised that winter would soon be upon us. Already the leaves were turning gold and brown and soon the trees would be bare. Remembering the rigours of last winter, I determined to be prepared this time and put Operation Siege into action. Soon the barn was bulging with bales of hay and sacks of oats, bran and maize. That took care of the donkeys and Barney. The big problem was the cats – they wouldn't eat tinned food and I only had a very small fridge. So I laid in a store of tins for myself and filled the fridge with rabbit, liver and fish for them. Drums of diesel and paraffin and cylinders of Calor gas were laid out in serried ranks behind the caravan, and comforting bottles of vodka and cartons of ciggies were tucked away in any spare corner I could find. I filled every available bucket with water, conscientiously refilling them with fresh water every day. I went round the caravan, hermetically sealing every gap and space. This done, I looked around and decided I was now ready for anything.

Gradually Simon was overcoming his nervousness and each day I was allowed to get just a little bit closer to him. The breakthrough came when I was brushing Humphrey. I noticed Simon's backside just two feet away and, almost without thinking, I reached out and very gently ran the brush over his bottom and down his nearside leg. He trembled and then, to my astonishment, moved about three inches closer. I tried another foray with the brush and he moved up another three inches. I didn't want to push my luck, so I carried on brushing Humphrey. Out of the corner of my eye I saw a shaggy behind gradually edging closer until it was practically under my nose. It really was too good an opportunity to miss so, making my apologies to Humphrey, I set to and started brushing Simon in earnest. To my delight he just stood there and let me get on with it. For over an hour I plied the brush and he ended up with what must have been the best-brushed bottom in the county. That was the snag: his backside was the only part of him I was allowed to touch. Any attempt to disentangle the hairy curtain over his eyes or on his back, and he was off. But I was encouraged that he always came back – always, alas, backside first. Even now, when we are the best of friends, I am only allowed to brush the rest of him as a concession.

After that we never looked back. I could approach him and he wouldn't move away. Soon he started making the first move, coming over and very gently nuzzling me. But what I found, and still do find, absolutely heart-rending are his pathetic little attempts at mischief. Quite clearly this isn't in his nature, as he is a rather sober and ponderous creature, but he seems to think it is expected of him. One of Humphrey's favourite tricks is to grab a coat-sleeve in his mouth and keep tugging away until coat and wearer part company. Simon can't bring himself to do anything quite as daring as this, but he very gently takes hold of your coat-sleeve with his teeth and gives tentative little tugs at it, almost as though he's trying to attract your attention, all the while gazing anxiously into your face for any sign that he's going too far. He still has a backside fixation. When I go out last thing at night with their peppermints and say, 'What about a goodnight kiss, then?' there is always one fat,

101

hairy donkey bottom among the soft muzzles pressed lovingly into my face!

Sitting on the loo one morning, I felt a gentle nibbling sensation in my toes. I looked down, and there was Pooh, almost bursting with self-satisfaction at having put one over on me. Investigating, I discovered that he had found a way into the caravan via a loose board under the bath – he was probably the one who loosened it in the first place. Having to concede defeat to a furry meatball isn't easy, but I know when I'm beaten. I told him he could have access to the bathroom but if he ever so much as set paw in any other part of the caravan he was out. He accepted this philosophically and set up home under the bath, industriously shredding the toilet roll to line his nest.

Having a resident ferret in the bathroom presented certain problems. I was never sure whether to warn

visitors about him or let them find out for themselves. If I told them, they wouldn't use the bathroom but sat cross-legged all evening, getting more and more distraught. On the other hand, if I kept quiet about it, there was no telling how they would react to finding that their most intimate activities were being shared with a ferret, however kindly disposed. In the event, my social life was considerably curtailed. There was also the problem of Charlie's rabbit foster-babies: I couldn't keep them in the bathroom now, not with even such a benign ferret as Pooh in residence. Fortunately, with winter coming on, the baby bunny market was declining, so I decided to worry about this problem when the time came. And I couldn't really blame Pooh for wanting to come inside for the winter – I wouldn't have fancied sleeping out in the barn myself.

The only discouraging thing was the cottage. I had hoped to move in by Christmas, but the way things were going, I would be lucky to be in by Christmas next year. It seems there are two kinds of people: those who have only to snap their fingers and workmen will drop whatever they are doing and run, panting with enthusiasm, to do their bidding; and those for whom the workmen happen to be working when the first kind snap their fingers. I fall into the second category. The builders were regularly disappearing, often for weeks at a time. They came back full of apologies and explanations about this very urgent job that had come up, they knew I wouldn't mind. 'How did they know I wouldn't mind?' I asked myself crossly. I *did* mind, very much. Presumably thinking they were being complimentary, they went on to explain that this other woman was such a pest and she kept on and on at them, making their lives a misery. She wasn't at all like me, they said. In the end they gave in, just to keep her quiet. This really infuriated me, because to get anything done, one apparently has to be bloody-minded and obnoxious. I was doing it all wrong – making them interminable cups of tea, baking them cakes because they refused to eat the shop variety and taking their ever-increasing defections in good part.

The architectural designer I had engaged to draw up the plans and keep an eye on the building progress turned out to be even more in awe of the workmen than I was. Every suggestion and criticism I put to him was met with shocked horror: 'Oh, I couldn't do that! Very touchy people, builders. Have to be handled with kid gloves, otherwise they'll walk out.' As they hardly ever seemed to be there anyway, I wondered how we would tell the difference. Meanwhile I was dishing out large sums of money in fees to the designer, marking time in a crowded and uncomfortable caravan while the builders vanished for weeks on end and having to put up with all sorts of deviations from the original plans which I didn't want but which both the designer and I were too cowardly to dispute. One thing I learned from all this – no way are the meek going to inherit the earth. Even if by some miracle they did inherit it, the non-meek, with elaborate explanations about this person they know who keeps on and on nagging them for it, would come along and take it off them. It all seems so unfair.

As it turned out, winter was a doddle that year, compared with the previous one. There was some snow but the path was only blocked for a few days. Humphrey and Barney grew thick coats and Simon looked more yak-like than ever. They have a spacious, weather-proof, straw-lined shelter in the paddock so I could never understand why they spent all their time outside in the rain, snow and biting wind – unless it was to make me feel guilty. I would go out with their food and they would be lined up by the fence, coats snow-spiked and eyelashes speckled with snowflakes, looking martyred and long-suffering as only animals (and particularly donkeys) can. They regarded me with their soft, melting eyes: 'You shouldn't have bothered, not for us. We don't mind standing out here in the freezing cold, catching our death. You go back to your nice warm caravan, we'll manage somehow' (deep sigh). I knew it was all an act but it got to me just the same. I would go back to the caravan beset with guilt and try to think up little treats to make it up to them – which, of course, was precisely what they intended. Quite unable to settle, I would be in and out of the caravan every five minutes bearing biscuits, sugar lumps, mints, carrots and similar goodies. They accepted these offerings graciously: 'If it makes you feel better, who are we to refuse?'

Quite the worst part of that winter, and every winter since, was keeping their water buckets filled and unfrozen. Several times a day I had to go out with containers of hot water to melt the ice. After a few days of drinking warm water, they decided they preferred it this way and ever since, even in the summer, I have had to top up their buckets with hot water to take the chill off. I don't know how the myth arose that man is the superior, intelligent, thinking animal. I now have six donkeys, and a good deal of my time is spent humping bales of hay and sacks of oats around. These activities are watched with great interest by the ultimate consumers, all of whom are considerably bigger, stronger and younger than me. It wouldn't occur to them to lend a hoof, and if I were to suggest it they would be most affronted: that's what people are for. So you tell me – which of us is the more intelligent?

Although it wasn't a bad winter, I wasn't sorry when it ended. Quite apart from the usual thrill and joys of spring, there was something extra to look forward to this year: Barney's lamb! My immediate inclination, as soon as the lambing season started, was to dash straight out and buy one. But my first year of country living had taught me some important, if mutually exclusive, lessons:

1. A new animal should never be introduced into the household until all necessary preparations have been made and all possible problems foreseen.

105

2. The foreseeable problems are not the ones that will crop up.
3. Nothing can be done about the unforeseeable problems so you might as well forget them.

With this in mind, I concentrated on the immediate problems, the most immediate of which was Barney. How was I going to keep him and the lamb apart until the lamb was big enough to take care of itself? One friendly nudge from Barney and bingo – instant shish kebab. There was only one thing for it: Barney would have to be tethered during the lamb-into-sheep transition period.

To get him used to the idea, I decided to start tethering him for a short while each day, gradually increasing the length of time so that when the lamb arrived he would take it in his stride. Put like that it sounds entirely logical and commonsensical, but would Barney see it that way? Riddled with guilt and misgiving, I made my way over to the paddock. I was halfway there when suddenly all hell broke loose: the donkeys started galloping madly round the paddock with Humphrey braying like a banshee, Barney was cantering up and down, bleating his head off, and even Henrietta, who almost never gives voice, was sitting on the fence tweeting excitedly (for such a large bird, she gives the most ridiculous little 'tweet'). I looked around to see what had brought on all this excitement and was amazed to see about fifty men in army camouflage gear skulking behind the paddock hedge. Good God, I thought, it's the Invasion – forty years late but they finally made it! I watched as they picked their way through the mud until they were level with the paddock gate and then, to my horror, they broke through the hedge into the paddock and started climbing out again over the gate, aided by an encouraging nudge on the rump from Barney. I grabbed hold of Humphrey before he had a chance to make a getaway through the gap so thoughtfully provided by these

cretinous vandals and tied him to the fence. The last of the mob, the officer in charge, was just being helped over the gate by Barney. 'Hey, you!' I yelled. 'Just hold it a minute.'

He turned round smartly, snapped to attention and saluted. I resisted the temptation to return the gesture with the two-fingered variety and, quivering with rage, asked him what in blazes he thought he was doing, breaking through my hedge with his tinpot toy soldiers.

Still standing to attention and looking straight over my head, he barked, 'Very sorry, ma'am. Only way through to path.'

'You blithering idiot!' I screamed. 'Why couldn't you have come through the paddock in the first place, using the gate like any other normal, civilised human being?'

'Very sorry, ma'am, couldn't do that. Manoeuvres. Mustn't be detected.'

I looked at him in utter disbelief. Undetected? With every animal in the place hysterically announcing their presence to the world at large? Suddenly the humour of the situation struck me and I just dissolved. 'You can't be serious,' I gasped. 'Undetected? With this lot?'

He looked discomfited. 'Orders, ma'am.'

'Yes, well,' I said, 'you'd better get your boy scouts back and mend this hedge before my counter-espionage agents get out and spread word of your movements to every farmer around.'

'Very good, ma'am.'

Now that my original rage had evaporated I was quite enjoying the situation. After several months of being treated like a witless moron by the builders and having my requests (I wouldn't dare give them instructions) completely ignored, I found all this saluting and 'yes ma'am-ing' quite exhilarating. I had to make a determined effort not to bark out peremptory orders: 'Check the fence while you're about it. Mend the catch on the gate. Fill the donkeys' buckets. Peel me a grape.' No doubt about it, power is heady stuff. Equally clearly, I was quite unfitted for it: a couple of days of this and it would be 'Eat your heart out, Idi Amin!' Over the next two years these 'secret' manoeuvres were repeated about once a month. The animals always gave me advance warning so that I could get out in time and enjoy my few brief moments of power. It did wonders for my morale.

I decided to postpone Barney's initiation into the joys of being tethered until the following day, on the grounds that he'd already had enough excitement for one day. In truth, I was glad of an excuse to put it off because I was absolutely dreading it. Next day I went into the paddock with a heavy heart. How on earth was I going to tell him that from now on the freedom he took for granted was going to be severely curtailed? He came barging over, full of trust and affection, which of course made me feel a thousand times worse. 'Come on, Barney,' I said, and led him out of the paddock into the garden. I gave him a cigarette and while he was happily munching, I fitted the collar and tied him to the stake – Heavens, it sounds just like Joan of Arc. There was no point in beating about the bush so rubbing his knobbly head, I explained that I hated this just as much as he did (or would, once the fact that he was tethered had penetrated) but it was for his own good. In time he would thank me for it. Listening to myself, I realised that I sounded exactly like the sort of well-meaning mum who brings me out in goose pimples. I cut my speech short, gave him another ciggy and left him to it. At five-minute intervals I went out to check that he hadn't tied himself up in knots or escaped and to reassure him that he was still loved. Each time he greeted me with a welcoming 'maa', which cut me to the quick. Why can't he snarl at me, I thought, or attack me? I wouldn't feel so bad about it then. But perversely, I wanted to feel bad about it, I deserved to feel bad about it. I wondered if Barney was smarter than I gave him credit for and this was his way of getting back at me. After all, the only really effective weapon animals have against us is their ability to load us with guilt. If that was his intention, he was succeeding admirably.

After half an hour I untied him and, anticipating a frenzied stampede, hopped smartly out of hoof's way. It was all a bit of an anti-climax, really: he just went on contentedly munching the grass around his feet.

Over the next two weeks I gradually increased the length of time he was tethered until he was eventually spending the best part of the day tied up. I couldn't believe the way he was taking this violation of his liberty; either he was too good-natured to object or too dim-witted to realise what was going on. Either way, I was enormously relieved that this run-up to Operation Lamb was going so smoothly. Time now for Operation Lamb itself. Round here it is mostly dairy and arable farming, so I had to go farther afield to find a sheep farm. I could have made enquiries locally, I suppose, but at this stage I didn't want word to get around that I was planning to introduce a sheep into the menage. After their experiences with Humphrey and Barney, I didn't imagine my neighbours would take too kindly to the prospect of yet another potential trouble-maker taking up residence within a mile or two of their property.

When I got to the sheep farm, production was in full swing. If I should ever meet that sheep farmer again face to face, I'm quite sure I wouldn't recognise him – but I would definitely recognise his backside! During the whole of our encounter he was down on his hands and knees, coaxing yet another little woolly bundle into the world, and his backside was the only part of him I got to know at all intimately. Feeling a bit of a fool, I explained my problem and asked whether by any chance he happened to have a lamb going spare. It seemed to me such an extraordinary request to make that I wasn't too sure how he'd take it. 'No problem,' he said, 'have three or four if you like.' He explained that some sheep produce more offspring than they have feeding stations for, others reject their babies and, tragically, some die in

childbirth. As the job of hand-rearing these orphans usually falls to the farmer's already hard-pressed wife, offers of adoption are enthusiastically welcomed. I was invited to take my pick of the foundlings, the only stipulation being that it had to be a baby ram (ramekin?). The farmer explained that as there was no end-product from rams, they were the more expendable. No male chauvinism among sheep farmers, that's for sure! I chose a speckle-faced, three-hour-old little moppet, purely on the basis of his speckles. Most lambs, like most sheep, look pretty much like most other lambs; at least Jason's speckles marked him out from the crowd.

I set off for home, stopping only to buy a baby's feeding bottle and a supply of the recommended brand of milk powder. Tiny lambs have to be fed every two hours, day and night, so by the time we got home Jason was ready for his feed. In fact I soon learned that whatever the time, Jason was always ready for his feed. Obsessive, like any other first-time mum, about the risk of germs, I boiled all the necessary equipment and sterilised my hands – boiling my hands, I decided, would be taking motherly concern a bit too far. Satisfied that no self-respecting germ would be found dead in such a sterile atmosphere, I mixed the feed and settled Jason on my lap. Before I even had a chance to offer him the bottle, he grabbed the teat in his mouth, clamped his gums round it and, tail wagging furiously, polished off the lot without once coming up for air. Appetite satisfied, he curled up on my lap and lapsed into unconsciousness. I was worried about wind. Should I burp him? He looked so peaceful that I hadn't the heart to disturb him.

Bedtime was planned like a military operation – better, in fact, remembering my earlier experience of a military operation. Four feeds were prepared and left in the fridge to see him through the night. Setting the alarm for two hours later, I tucked Jason up in his basket, complete with blanket and hot-water bottle (my hot-water bottle – I know of no greater sacrifice) and installed him by the side of the bed. Charlie watched all these preparations with considerable interest. He was agog with curiosity about this newcomer and was dying to get his paws, or rather his tongue, on him. I was tempted because, after all, sheep lick their offspring and as this was one maternal chore I didn't feel inclined to take on myself, why not let Charlie do it? Also, he would provide warmth and comfort for this motherless waif. But what about germs? Thinking about sheep I have known, admittedly not all that intimately, I had to concede that they didn't look all that hygienic. Quite mucky, in fact. I decided to take a chance. Charlie was delighted and, easing himself gently into the basket, he clasped Jason firmly between his front paws and treated him to his orphan's special (ovine variety) once-over-lightly. Jason bleated happily and, tail going like the clappers, immediately started searching for a feeding station. Embarrassed, Charlie looked the other way and pretended it wasn't happening.

Pudding, Floss and Min, after a few desultory sniffs, decided that this funny-looking woolly cat wasn't interesting enough to bother about and thereafter completely ignored him. Rufus, however, was put out. His 'take-no-notice-and-it-will-go-away' routine wasn't working – far from going away, these interlopers just kept on coming. It was too much. He decided that more positive measures were called for and assumed his well-practised long-suffering martyr rôle: 'Not another one to replace me in your affections? Please don't worry about me, I'll just sit quietly by myself, just me and my memories. I know I'm a burden to you but it won't be for much longer.' It was pure ham. He knew it and I knew it, but it got to me just the same. Anguished, I tried to think of ways to make it up to him. I fussed over him, spoiled him outrageously, even made up a feed for him every time I fed Jason so that he wouldn't feel left out in the cold. He graciously accepted all these offerings, making

111

it clear he was only doing it 'so that you'll have nothing to reproach yourself with when I'm gone.'

As it turned out, I needn't have bothered about the alarm. Exactly one hour and fifty-five minutes later I was woken from a deep sleep by a muted whimpering in my left ear. By the time I had levered my second eye open, the whimpering had given way to a persistent bleating. I hopped out of bed and warmed Jason's bottle to the accompaniment of an ear-blasting, full-blooded 'Maagghh!' I couldn't believe that so much noise could come from such a tiny scrap of a thing. I hastily put the teat into his mouth and the caterwauling subsided into a contented gurgling. He finished the bottle in about three minutes flat and two seconds later was fast asleep again. I refilled the hot-water bottle and changed his bedding – tiny lambs are very, well, loose. Charlie, after cleaning himself thoroughly from whiskers to tail, decided that incontinent lambs were probably better off sleeping alone and reverted to his usual bedtime position – on my face.

For the first two weeks or so Jason was absolutely no trouble at all. He ate and he slept, and that was his life. As soon as he woke up he ate, and as soon as he'd finished feeding he slept. Really, I thought, I can't understand why mothers with young babies make such a fuss; there's nothing to this baby-rearing lark. Come-uppance time was just around the corner. Jason was growing fast and developing a will of his own. What's more, it was stronger than mine. If I wanted him to do something he didn't want to do or, more often, if he wanted to do something that I didn't want him to do, there was no contest – he always won. He decided that a basket with a hot-water bottle was all very well, but a bed with a resident mum was a whole lot better. He started climbing into bed with me; if it was good enough for the cats, it was certainly good enough for him. I pointed out to him that for one thing, the cats were house-trained and he wasn't even bed-trained. Also, cats have soft paws, not rock-hard hooves, and a loving pat on the nose from the cats did not result in a livid, agonising bruise. Lesson number one: you can't reason with a lamb. He listened intently to all my arguments, nodding sympathetically from time to time and, immediately I had finished, hopped straight back onto the bed again. I began to dread bedtime. Fit to drop after the day's activities and suffering badly from loss of sleep, I was in no condition to start fighting a territorial battle. Reflecting that I must be a real pushover if even a pint-sized lamb could put one over on me, I eventually gave in. It was less exhausting. But I did take the precaution of togging Jason up with protective clothing (my protection, not his): thick bedsocks and a pair of voluminous drawers!

He then decided that sleeping during the day was a waste of time, time that could be much better spent falling into water buckets, jumping onto hot stoves, tripping over cats and a thousand and one other potentially suicidal activities. I didn't dare take my eyes off him or leave him alone for a minute unless I bound him hand and hoof – and don't think I didn't consider it! Jason for his part watched me like a hawk to make sure I didn't go anywhere or do anything without him. I couldn't even go to the loo unaccompanied but he came skipping after me, 'maa'-ing his silly little head off. 'Now look,' I said to him, 'I am not your maa. Your ma is a fat, dumpy, woolly lady.' He pondered this for a moment, looked me over from top to toe and then said even more emphatically 'Maa.' Well, thanks, I thought.

Because of his feeding schedule, not to mention his apparent death wish, I had to take Jason with me wherever I went. He thought this was great fun: driving around in the car, going shopping on a lead, fawning obscenely over the besotted passers-by who stopped to pat him and taking in extremely good part the highly personal comments on his appearance – 'Funny-looking poodle' was the most usual. He was an absolute knockout with traffic wardens; rubbing his woolly little head

113

against their legs and gazing adoringly into their eyes, he managed to push the matter of excess time on the meter right out of their minds. Even world-weary traffic cops fell under his spell. By the time they'd got over the shock of finding a lap-lamb in the car and had made complete fools of themselves by clucking idiotically to him, they usually forgot what they'd stopped me for in the first place. You can keep your mechanical nodding dogs – give me a real live lamb any time.

Soon Jason was big enough to join us on our afternoon walks and he certainly added an extra dimension to them. He loved these outings, frisking and cavorting with an exuberance that made me feel like a dinosaur. He never entirely gave up his gambolling, even when he was fully grown. I had only to say to him, 'Come on, Jason, do a frolic for me,' and he would heave his fat, solid bulk into the air, perform a ponderous gambol and drop like a lead balloon. He chased rabbits, played silly-lambs with the cats and hide-and-seek with Pooh. From time to time he would suddenly take off at the gallop and, when he was no more than a dot on the horizon, realise that he couldn't see me any more. Skidding to a halt, he did a nifty U-turn and came charging back. He never applied his brakes on this return journey, preferring to use me as a buffer. Like Barney, he was always very sorry about it, planting both front hooves solidly on my chest as I lay on my back and gasped for breath, and 'maa'-ing anxiously in my face to explain how very sorry he was. 'OK, Woollychops,' I said. 'If you're big enough to knock me down, you're big enough to take your chances with Barney.' True, he couldn't quite match Barney for height, but by this time he certainly out-did him in bulk. He was also quite a bit stronger than Barney, who himself was no weakling, so I had no misgivings about the meeting. Jason could more than take care of himself if it came to a confrontation.

Bubbling with excitement, I led Jason over to where Barney was tethered. I felt just like Father Christmas bearing goodies. 'Look, Barney, a friend for you!' There's nothing like an animal for putting you down. Barney looked up from his grazing, fixed Jason with a contemplative stare and, plainly completely underwhelmed, carried on eating. Jason wandered over to Barney and gave him a tentative sniff. Barney, his mind clearly on other things, half-heartedly sniffed back. Feeling no end of a fool, I consoled myself with the thought that at least they hadn't attacked each other and that, really, it was much better this way. All the best friendships started out on a low key and developed from there. I untied Barney and after a few more cautious sniffs, the pair of them set off on a sedate stroll around the garden. I was deeply touched. Now that Barney had a companion, I was convinced that all my troubles were over. I never learn. In fact, my troubles hadn't even started yet.

115

CHAPTER 13

For the moment, however, everything was going smoothly. Jason and Barney were happy together, Humphrey and Simon had each other, the cats were thriving. Fat little Pooh was bustling around, industriously rebuilding his nest and costing me a fortune in toilet rolls. I had a sudden thought about all this nest-building activity: could it be that spring had got to him, too, and he was getting broody? The fact that I always thought of Pooh as 'he' didn't mean a thing; in the absence of any evidence to the contrary, to me all animals are 'he' until proven otherwise (sorry, Germaine Greer). I went into the bathroom and called him – he answered to the name of 'grub's up'. He came wriggling out from under the bath, squeaking excitedly. I picked him up and examined him. He did seem to be putting on a bit of weight. Was it possible that he could be in a Certain Condition? I don't know a lot about ferrets but I thought it unlikely that they reproduced by virgin birth. So, if my suspicions were correct, another ferret must have been involved. I was incensed. This pushy little upstart had been carrying on right under my nose. I fixed him with my 'I don't wish to play the heavy parent but I insist on knowing what's been going on under my roof' eye. He looked coyly back at me. Until then I hadn't got beyond the question of Pooh's unspeakable behaviour, and the matter of end-products hadn't crossed my mind. Suddenly the full implications of the situation struck me: dozens of baby ferrets, all living under my bath, ponging like mad and, unlike Pooh who was definitely a one-off, probably mean and wild and lusting for blood. I couldn't bear to think about it, so I decided not to.

Casting around for something else to worry about, I settled on Henrietta. She was lonely, I decided. She needed a mate or, at the very least, a companion. There could be no question of getting her a mate in the true sense of the word; for one thing, her ancestry was so dubious that the chances of finding a sexually compatible mate were so remote as to be non-existent, and anyway she was too old for that sort of nonsense. But a soul-mate, somebody to bring her comfort in her declining years – surely that was the least I could do for such a sweet old lady? In view of her alleged parentage, the obvious choice was a peacock or a guinea fowl. I couldn't really afford a peacock and in any case I wouldn't have the nerve to introduce one of those aristocratic creatures into the boggy, weed-infested slum I euphemistically call the garden. Also, if I'm honest, I wasn't sure that my ego was up to it: it had taken enough of a bashing what with donkeys treating me like dirt, lambs putting me down and ferrets cocking a snook at me, so the idea of being sneered at by a supercilious peacock was more than I could stand. It would have to be a guinea fowl.

Of course I came up against Sod's Law again: 'A guinea fowl? At this time of year? Oh no, me dear, 'tis the wrong time for 'em.' Even worse were the near-misses: 'Ah, now, if you'd only asked me last week/two days ago/yesterday, I had just the thing for you. Lovely bird, hand-reared, going for a song.' Oh, sure. But I met some delightful people during my quest, in particular a breeder who had the most fascinating collection of birds imaginable. I can't even remember now how I came to hear about him. I drove into the yard and as I got out of the car a voice said, 'Hallo there. What do you want?' I looked around but there wasn't a soul to be seen. Feeling a bit of a fool, I said, 'I'm looking for Mr Newbury.'

'Oh, he's gone, he's gone. He's not here, he's gone.'

Just then a tall bearded man came out of the house. Laughing, he said, 'Take no notice of him, he's a shocking liar,' and introduced me to his receptionist – a jay sitting on the gate post. Intrigued, I asked Mr Newbury how he had taught the bird to speak. 'Oh I didn't teach him, he just picks it up from people around the place.'

Mr Newbury didn't have a guinea fowl, not at the moment. But he did have a pair of the most enchanting Chinese goslings about three weeks old, still covered in their baby down but already showing signs of the long-necked elegance to come. I wanted them desperately but couldn't think of any justification for buying them. Impulse purchasing of animals just wasn't on, I told myself. Once I started that sort of nonsense the whole thing could get completely out of hand. As if reading my mind, Mr Newbury came to the rescue. 'Make marvellous guard dogs, geese.'

Of course! Exactly what I needed! They'd keep predators away, warn me of intruders, protect me from prowlers – in short, do everything a dog would do without actually being a dog. Well-meaning people were always telling me, 'What you need is a dog.' They were right, of course. They knew it and I knew it, but the problem was getting the cats to see it. I had twice tried to introduce a dog into the household and on both occasions they had left home, making it quite clear that they had no intention of setting paw in the place again until that Thing had gone.

Gussie and Griselda immediately adopted me as their mum. As soon as they caught sight of me they came hurtling over, honking with delight. I was rather hoping that Henrietta might take them to her bosom: it would solve her loneliness problem (which probably existed only in my mind) and at the same time provide the goslings with a much more suitable, feathered foster-mum. She did show some interest at first, tripping fastidiously over to see what all the fuss was about. She watched the goslings' antics impassively for a moment or two, decided that these ill-mannered yobs were definitely not the sort of person one would wish to associate with and, with a contemptuous toss of her head, swept majestically off.

The ground in front of the cottage was criss-crossed with ruts left by the builders' lorries. Watching the goslings happily splashing around in one of these rain-filled ruts one day, I thought it would be a nice gesture to enlarge it a bit so that they'd have more space to slosh about in. I got the spade and started digging. I don't quite know how it happened, but I got completely carried away and they ended up with a pond some ten feet across. They were delighted with it. On an impulse, I decided to give the toads their freedom. Perhaps they would breed in the pond and present me with hundreds of little toadlets. I went inside and rounded them up and tucked them under a mossy stone by the side of the pond.

Although the goslings spent their days in the garden, it was still a bit too chilly for them to sleep outside so I brought them in every evening. As soon as I sat down to watch television they were all over my feet, squawking to be picked up. I put them on my lap, and after much squabbling and jockeying for position, they settled down happily to watch the programme. They turned out to be the sort of viewer who can't watch anything without commenting noisily throughout – the kind of person I always seem to sit next to at the cinema. Every word was echoed by a joyous honking which completely drowned out the rest of the dialogue. After a couple of evenings of this I gave up even trying to watch; I switched on the set for them and buried myself in a book.

Looking back on those happy early days with Gussie and Griselda, it is difficult to pinpoint exactly when things changed. All I know is that from being affectionate,

119

gentle little creatures, the goslings turned into aggressive, bad-tempered bullies. Of course, that was why I had got them, because guard geese have to be aggressive. The trouble was that my guard geese were only aggressive towards me. Complete strangers could come along the path and walk right past them and they never gave so much as a honk, but immediately they caught sight of me they came charging over, necks low along the ground and hissing like cobras, vying with one another to see who could get the first bite. I never dared turn my back on them; twice they crept up on me from behind and pitched me headlong into the pond, on the second occasion sitting on my head so that I almost drowned. I tried to think back over their babyhood to see if I could find a clue to their behaviour. Had I been too strict with them? Scolded them too severely? Deprived them of a worm? Refused to let them watch the television programme of their choice? I could find nothing and was forced to the unhappy conclusion that once they reached the age of discrimination they simply decided that they didn't like me. And to think that I hadn't wanted to get a peacock in case my fragile ego was bruised!

On the whole, though, things were going very well. I should have known that it couldn't last, that there had to be a cloud behind the silver lining. Two clouds in fact – Barney and Jason. It took Jason no time at all to devise his own method of breaking out of the paddock. Too fat to squeeze through gaps in the hedge like Humphrey and too heavy to vault it like Barney, he simply leaned against the fence until it collapsed under the strain. Of course Humphrey, never one to miss an opportunity, gleefully took advantage of these new escape routes opened up by Jason. As for Barney, he regarded Jason not so much as a companion as a disciple. Every wicked trick he knew he taught to Jason. Jason, catching on fast, soon started thinking up tricks of his own which, on the principle that one good turn deserves another, he generously passed on to Barney.

Naïvely, I had hoped that now that Barney and Jason had each other they would make fewer demands on me, but it didn't work out that way. Clearly concerned that I shouldn't feel excluded, they went to considerable pains to let me know that I was still wanted. I now had double trouble – two busybodies involving themselves in what-ever I was doing instead of one, two pairs of eyes watching my every activity and two pairs of ears picking up any sound of a move being made without them. It had been hard enough to out-manoeuvre them individually, but as a team they were invincible. The moment I put a foot outside the door they came charging across like a pair of demented buffaloes, scattering everything in their path. Firm believers in the precept that the shortest distance between two points is a straight line, they never went round anything if they could go through it. Plants, tubs, water buckets, garden seats – everything went flying as they crashed their way through. My ribs were a mass of bruises from the double buffeting they were now subjected to; there was no way I could avoid the onslaught, so I did the next best thing and threw myself flat on the ground before they could do it for me.

Soon they started wandering farther afield, leaving a trail of destruction behind them, trampling crops, eating carefully tended gardens, uprooting young trees, fright-ening the life out of chickens and ducks. Every time the phone rang, my heart missed a beat, terrified that it was yet another complaint. It usually was. I was at my wits' ends and just didn't know what to do. I couldn't bear the thought of keeping Barney and Jason tethered for the rest of their lives, but on the other hand I couldn't let them carry on devastating my neighbours' property. Then fate, or whatever you call it, took a hand. For some time I had had contact with a farm, run as a youth venture, in north London. I was always very impressed by the way the animals were cared for and the obvious love and attention they received. So when the superintendent offered to take in my two delinquents, I knew this was

121

the answer. I was heart-broken at having to part with them but I knew that they would be well looked after and also that they would never be short of human companionship as there were always visitors to chat them up, give them treats and generally spoil them.

I always make a point of going to see them when I am in London and I am always reassured to see how well and happy they look. Except on one visit, when I noticed that Jason had his leg in plaster. It seems that he and Barney had got into the habit of strolling down the road to the pub every evening, where admirers would stand them the odd glass of beer. On the way back from one of these jaunts, and probably a bit unsteady on his pins, Jason was hit a glancing blow by a taxi. Fortunately there were no complications and the leg healed perfectly, but I have often thought about that taxi-driver and tried to picture the scene when he arrived home that evening and explained to his wife, 'Sorry I'm late, dear, but I knocked down a sheep in Kentish Town.' Well, would you believe it?

On one of my visits to the youth venture farm, I found myself being accompanied by a tiny bantam cockerel. Wherever I went, he followed. He really was the most exquisite creature, with long, silky feathers of every hue – peacock blue, emerald green, dark red – and with a long plume of blue-black feathers curling over his back. Every time I stopped, he stopped too, looking up at me with a quizzical eye. I sat down on a box and said, 'All right, then. What's it all about?' He hopped onto my lap and said 'Aarrk.'

'I see,' I said, 'you just want somebody to chat to, is that it?' He nodded his head vigorously and said 'Aarrk, aarrk.' I was a bit nonplussed – what on earth does one talk to a cockerel about? My experience with domestic poultry was limited to eating the fruits of their labours. On the premise that a little bit of flattery never comes amiss, I told him that he was quite the most beautiful bird I had ever seen and that I would dearly love to own him. To my intense embarrassment, the superintendent, who must have been listening to this idiotic one-sided dialogue, suddenly appeared as if from nowhere and asked, 'Would you really like him? We're trying to find a home for him.' I couldn't believe it. Why would they want to give such an enchanting creature away? With much humming and hawing, the story eventually came out. Reduced to its basics, it seemed that the little cockerel was Not Quite Normal – in other words, he was thought to be gay. The other cockerels made his life a misery and the hens despised him, which was why he tended to attach himself to people. Always a sucker for the underdog, I said that of course I'd love to have him, and bundled him into the car before they had a chance to change their minds. I christened him Joseph because of his coat of many colours, but it didn't really suit him so he became Josh.

Josh was the most endearing little chap, full of love and affection. He always accompanied me on my rounds of the other animals, sitting on my shoulder and clucking happily in my ear. At night he roosted in the apple tree but well away from Henrietta, of whom he was clearly very much in awe. Henrietta for her part completely ignored him. At first light he was on the window sill, tapping away with his beak in Morse: 'Time to get up. Breakfast time.' It says a lot for his charisma that the thought of wringing his neck for waking me up at such an ungodly hour never even crossed my mind. But soon that old guilt feeling came creeping in. Much as I loved having him trotting around behind me, and much as he obviously enjoyed it, I couldn't get over this niggling feeling that it wasn't natural. He should have companions of his own kind. But where does one find a gay cockerel? I thought about this for some time. Is he really gay, I wondered? Perhaps the other cockerels picked on him because he was so beautiful and they were jealous, and the hens despised him because he was too good-natured to stand up for himself? Perhaps if he had his own harem, with no other cockerels to challenge his position as top cock, he would find his real identity.

Ever since I moved to the country, people have been saying, 'You really should keep chickens.' Even my London friends, whose ignorance of country life in no way deters them from laying down the law on matters they know absolutely nothing about, were forever going on about the joys of poultry-rearing and free-range eggs. I took no notice, because nothing in the world would induce me to keep chickens. They were messy, stupid, totally lacking in personality – no, thank you. My relationship with Josh in no way altered my feelings about chickens in general because he, like Pooh, was a one-off. Then I sat down and examined my motives. All right, I didn't like chickens. But were my feelings paramount? Having taken on the responsibility of Josh, I owed it to him to put his needs first. If he needed a clutch of wives, then he should have them. Adjusting my halo, I trudged across the fields to have a word with Alec, who keeps a wide variety of poultry and is very knowledgeable about these things. He very generously agreed to let me have five bantam pullets. Within a few days they were installed in a chicken-house hastily cobbled up from one of those auction sideboards. I explained to them that they were only expected to sleep and lay their eggs in this chicken shack, and the rest of the time they could range free. They cocked a collectively beady eye at me: 'You do it your way and we'll do it ours. OK?'

The change in Josh was astounding. He puffed out his chest and immediately got to work showing the chickens who was boss around here. One look at him and they swooned in ecstasy. It took him less than ten minutes to prove that rumours about his predilections and virility were totally unfounded. His wives were delighted and so was I.

Over the next few weeks I learned a lot about chickens. I learned that, given a warm, cosy, hay-lined nesting-box to lay their eggs in, they will opt for a muddy ditch, under the van or in the donkey-house – in other words, anywhere so long as it's totally inaccessible. In the same way, they will scorn a comfortable, secure roosting-house and spend their nights on the lowest, most

fox-accessible branches of a tree, frozen stiff and soaked to the skin and adding considerably to my burden of guilt. I also learned that this delightful business of sliding your hand underneath a sitting hen and extracting a warm, freshly laid breakfast egg is just a myth. For one thing, your average hen will strongly resent this invasion of her privacy and will make her objections painfully clear. For another, the chances are that all you extract will be a handful of extremely messy chicken excreta.

I spent the best part of each day searching for their eggs and of course, once I discovered a nesting-place, they immediately stopped laying there and found somewhere else. It turned into a ludicrous game of hide-and-seek, with me always the loser. In the end, to my everlasting shame, I was reduced to spying on them. I hid behind a tree, armed with a pair of binoculars, and monitored their movements, pinpointing the spot where they disappeared and sneaking off after them. They were always one step ahead of me, never moving in a straight line but taking a highly convoluted route and frequently back-tracking to put me off the scent. They always managed to outwit me, and I was forced to re-think my idea of chickens as stupid and unintelligent. If they were stupid, what did that make me?

After a time I also found myself re-adjusting my ideas about chickens' lack of personality, because they were definitely emerging as quite distinct characters. Mary was the largest, a white, very pushy lady and undisputed boss. Nobody else was allowed to eat until she'd had her fill, and sometimes not even then. Her second-in-command was Dinah. When Mary was busy laying her egg or fulfilling her wifely duties to Josh, Dinah was allowed to take over, on the strict understanding that this was a purely temporary arrangement. Then there was Clara, a buff-coloured matron with an IQ of approximately three and a half. I never had any difficulty finding Clara's eggs, because she always laid them on the sloping roof of the barn and they were very easy to spot by the broken shells and gooey mess on the ground below. Last in the pecking order were Castor and Pollux; they were Polish twins, black and gold with enchanting little pom-poms on their top-knots. They worshipped Josh from afar but were never allowed to consummate their passion if Mary was around – Josh was too frightened of Mary to do anything that might incur her displeasure.

I was thrilled with my new family, and I was also glad to have something to take my mind off Barney and Jason. I was missing them dreadfully.

CHAPTER 14

'Jenny donkey and foal, free to good home.' I read the ad again to make sure I'd got it right. I couldn't believe that people were actually giving donkeys away. And to think of all the hassle I'd had, trying to find a companion for Humphrey! At odd times during the day I found myself thinking about those two donkeys. What sort of person gives away their donkeys? Even worse, what sort of person takes them? Someone who wants something for nothing? I thought of the story about the dealer trying to sell an elephant to a man in a pub. The more the man protested that he had no room for an elephant, he lived in a flat, he was out at work all day, his wife hated animals, the more pressing the dealer became. Finally he said, 'Look, I'll tell you what I'll do. As a special favour, you can have three elephants for the price of two.' 'In that case,' said the first man, 'I'll take six.' I had visions of those donkeys going to someone who didn't really want them but couldn't resist a bargain. They'd end up in a back garden somewhere as a nine-day wonder, and when the novelty wore off and the price of hay and oats dawned, they would become just two more candidates for the animal welfare societies – if they were lucky. If they were unlucky they'd be turned into hamburgers. True, the ad did say 'to good home', but would the advertiser bother to check? It seemed to me that anybody who was casual enough to give away their donkeys via a small ad in the local paper wouldn't be too bothered about the kind of home they went to.

I tried to put it out of my mind but it was no good. In the end I wrote to the advertiser, pointing out the dangers of offering free donkeys to complete strangers. Let her call me an interfering old busybody, I thought – everybody else does. That was really all I meant to do, so I was surprised to find myself adding that, if no really suitable offer came along, I would be happy to take the donkeys myself. That was that – or so I thought until I got a phone call a few days later thanking me for my kind offer, and when could I collect the donkeys?

Suddenly the full realisation of what I had done hit me and I went weak at the knees. Just when everything was going so well, when Humphrey was settling down with Simon and the neighbours were talking to me again, I was going out of my way to introduce a spanner, or rather two spanners, into the works. Up to now, my main concern had been the fact that these donkeys were being given away; now, for the first time, I began to wonder why. Nobody gives away nice, gentle, well-mannered donkeys, only bolshie, bloody-minded ones. And I was planning to introduce this pair of horrors into my now reasonably well-ordered menage. Had I gone quite mad? Undoubtedly.

Full of apprehension, I set off to collect the donkeys. The effusive welcome I got from their owner did nothing to reassure me. She was *so* pleased to see me, *so* glad I hadn't had second thoughts. Don't kid yourself, I thought, I've not only had second thoughts, but third, fourth and fifth as well. The fact that she didn't ask me anything about my experience with donkeys or the kind of accommodation I had for them confirmed my suspicions that she wasn't too bothered about their fate. For all she knew, I could have been a horsemeat-dealer. She went off to get the donkeys while I passed the time gnawing my fingernails down to the elbows. By now I was so screwed up about the whole business that if she had led out a couple of rabid, foaming-at-the-mouth bucking broncos I would have been quite prepared. So the slim, shy grey Jenny and her chocolate-brown imp of a son took me completely by surprise. They stood quietly by whilst the introductions were made, gently nuzzling my hands by way of greeting. I was enchanted – and relieved, not only because I hadn't clobbered myself with a pair of monsters but also because these two charmers had been saved from a possible fate worse than death. As the titles were now vacant, I christened them Phil and Sophie.

I still had the biggest hurdle of all to face: introducing them to the incumbents. I hadn't mentioned the matter to Humphrey or Simon because the less time they had to brood about it and plan reprisals, the better. Not that I was too worried about Simon, for I knew him well enough to be confident that he would accept the newcomers, as he accepted everything else in his life, with his custom-

ary tolerance and good nature. But Humphrey – this was a horse, or rather a donkey, of a different colour entirely. Still, there was no point in anticipating trouble. I'd just have to play it by ear. I backed the trailer into the paddock and let down the back. Humphrey immediately came charging over wearing his 'What's all this about, then?' expression. He didn't know what was in the trailer but whatever it was, he was not going to have it. Just as I was expecting all hell to break loose, Sophie came mincing daintily down the ramp. She sidled up to Humphrey, fluttering her incredibly long eyelashes, and simpered coquettishly at him. The effect this had on Humphrey was electrifying. All the belligerence went out of him and, metaphorically straightening his tie and smoothing his hair, he leered suggestively at her. I was shocked to my vitals: clearly there were depths to Humphrey still unplumbed. Sophie, eyes glowing with adoration, gently nibbled his neck and he all but purred. 'Nice one, Sophie,' I told her, 'I like your style.'

That was one problem out of the way. We now had a new one. How was Simon going to feel about sharing Humphrey with Another? I looked around for him; to my utter astonishment, he was in the trailer, carefully grooming little Phil from top to toe. Toilette completed, he gently nudged Phil down the ramp and took him on a tour of inspection round the paddock. I couldn't believe the way this was working out – it had to be too good to last. Right from the start, Simon appointed himself Phil's nanny, protector and guardian. Watching the two of them together, it occurred to me that Simon really was everyone's idea of a favourite uncle: large, shaggy and bumbling, kind and generous to a fault, firm when firmness was called for but essentially gentle and loving. Lucky Phil, I thought. Sophie took this infringement of her maternal rights in surprisingly good part. Once she got over the initial shock, she realised that she was on to a good thing here – a ready-made baby-minder virtually on tap. Casting aside the cares of motherhood, she set

131

her cap at Humphrey. Flirting outrageously and throwing modesty to the wind, she used every trick in the book to win his favours. 'You don't have to go overboard,' I told her. 'He's a pushover. One worshipful glance and he's yours.' She eyed me coldly: 'What would you know about catching a man?' Maliciously, I thought, 'I know things about Humphrey you don't know. For one thing, he's Incapable.' As if reading my mind, Humphrey shot me an anxious glance. 'Oh, all right,' I said, 'don't worry. She won't hear it from me. But remember you owe me one!'

Simon took his responsibilities as Phil's guardian very seriously indeed. He watched over him constantly, never letting him out of his sight for a moment. The slightest hint of danger and Phil would find himself being hustled firmly into the donkey-shelter. As pretty well everything came under the heading of danger so far as Simon was concerned, poor Phil was in and out of the shelter like a souped-up shuttlecock. The other animals were as surprised at this sudden change in Simon as I was, because they had always regarded him as a soft touch. Accustomed to going in and out of the paddock as they pleased, they now found it was strictly out of bounds. There was nothing pushy or aggressive about Simon's attitude; he just made it quite clear that in future any unauthorised creature wanting to enter the paddock would have to get past him first. They took one look at his solid bulk and decided not to bother – all except Charlie, that is. Nobody tells him what to do. He simply climbed onto the fence and jumped over this hairy obstacle.

As for people, they represented the biggest threat of all. An unfamiliar voice or the sight of a rambler on the horizon and it was Action Stations. Pushing Phil unceremoniously into the shelter, Simon would take up sentry duty at the entrance. Considering that Simon himself was still very nervous of strangers, I was immensely touched by his courage in defence of another donkey. Meanwhile Phil was growing fast and he was beginning

132

to find Simon's attentions a bit of a drag. Apart from meal-breaks, when Simon reluctantly relinquished him to Sophie, his every move and action was monitored by this large, shaggy busy-body. He loved Simon, of course, but well, he wasn't much fun. Now Humphrey, on the other hand . . . My heart sank. For some time now I had noticed, but refused to recognise, an all-too-familiar gleam in Phil's eye, a spark of mischief that spelled trouble. A touch of the Humphreys, in fact. 'Dear God,' I pleaded, 'let me be wrong. Not another Humphrey, please.' As it turned out, God wasn't listening.

Still, all that was in the future, and right now I had a more immediate problem on my hands. I always looked forward to Pooh's greeting every morning when I went into the bathroom. He came wriggling out from under the bath, his fat little body rippling with excitement. Rolling over onto his back, little legs kicking in the air, he waited for his morning tickle, squeaking joyously. But for two days there had been no sign of him and I was out of my mind with worry. Perhaps he's had his babies, I thought (I still thought of him as 'he'). I peered under the bath with a torch. His toilet-paper nest was there, but no Pooh and no babies. I walked round the garden and paddock, calling and calling. Nothing. Then, on the second day, Charlie suddenly shot off across the garden and a few minutes later I heard him calling excitedly. I dashed over and saw his rear end sticking out of one of the pipes stacked behind the cottage by the builders. One of Pooh's favourite games was to hide inside a pipe and then jump out at me as I passed. Fearing the worst, I hauled Charlie out and looked inside. There was poor little Pooh, firmly wedged and past all help. Clearly he hadn't reckoned on his expanding waist-line but must have squeezed in and got stuck. I manoeuvred him out and buried him under the apple tree. Charlie watched soberly. He was going to miss his playmate. I scratched his ears: 'I know, Puss,' I said, 'but it's going to come to all of us some day.'

I went about the chores with a heavy heart. 'It's ridiculous to get so upset about a ferret,' I told myself. But he wasn't just a ferret – he was Pooh. I watered the donkeys, engaged in the usual battle with the geese and was feeding the chickens when I noticed that Dinah was missing. 'All right, you lot,' I said, 'where is she?'

'Don't look at us,' they replied, 'we've got our own problems.' And off they went to hide their eggs. I could see it was going to be one of those days.

As day succeeded day, with still no sign of Dinah, I was forced to accept the inevitable, that she had been fox-napped. So when, some three weeks later, she suddenly emerged from under the donkey-shelter, closely followed by thirteen little pom-poms on legs, I was overjoyed. They were the first babies to be born since I had moved in. Shortly after, Mary followed suit, proudly presenting me with fourteen balls of fluff – of course, she'd have to be one up on Dinah. By the time  Castor and Pollux had produced nineteen chicklets between them, I have to admit my joy was less than unconfined. I was pretty well ankle-deep in baby chicks, all eating their heads off; meanwhile I was having to buy my eggs from Alec. Economically speaking, there was something very wrong somewhere. It was then that I came face to face with Poultry Keeper's Law number one – out of every ten chicks hatched, seven will be cockerels. So in no time at all I was playing host to twenty-eight cockerels, all with enormous appetites and no potential end-product apart from copious amounts of chicken manure, for which there is a limited market.

I sat down and did my sums. Taking into account the cost of chicken feed (which, believe me, is *not* chicken feed), bedding and chick crumbs for the babies, and completely ignoring the labour involved, I worked out that each egg produced cost me something over one pound. And as the baby chicks' diet was supplemented with chopped hard-boiled egg, it didn't take a mathematical genius to work out that egg production was not exactly an economic proposition. At least, not the way I did it. Apart from the cost, I couldn't keep all these cockerels because Josh, very much the proud father when the baby chicks emerged, was now getting hag-ridden by the non-stop battle for top cock among his male offspring. Basically a pacifist, he threw in the sponge and took to following me around again. Quite clearly, things couldn't go on like this.

As I parked myself on the garden seat to ponder, Josh hopped onto my lap and said 'Aaark.' 'I know,' I said, 'we've both got a problem. But don't worry, we'll think of something.' Yes, but what? My farmer neighbours, unhampered by sentiment, were in no doubt about the solution: 'Eat the buggers.' I was appalled. Eat them? I could no more eat those chickens than I could my own mother. I knew them all by name, had watched them growing up and developing into distinct personalities. Little Ferdy, who thought he was a duck and had to be fished out of the pond at least twice a day. Horace, with his lop-sided comb and pathological terror of worms. Bertie, who had fallen madly in love with Griselda and was regularly beaten up by Gussie for his pains. Even Harold, who absolutely loathed me and spent his days lurking in unlikely places for the sheer joy of jumping out on me as I passed. They were all family – and you don't pluck, stuff and eat your own family. In the end I took the coward's way out and gave them to Alec. I didn't ask what he did with them and he didn't tell me. I preferred not to know.

Foxes were taking their toll, too, and I was finding it difficult to reconcile my admiration and respect for them with the depredations they were making on my birds. I was in the donkey-house one morning when I suddenly heard Henrietta tweeting hysterically. I rushed out and saw her in the middle of the paddock, running around in circles while a fox crept up on her. Why she didn't take flight, I don't know. The fox was within fifteen feet of her and getting closer all the time. I dashed over, screaming

'Get out, go away, shoo!' The fox just kept on coming. I picked up the first thing that came to hand, which happened to be a lump of donkey dung, and threw it. Unfortunately it hit Henrietta, but it had the effect intended and the fox streaked off into the woods. I was devastated: what on earth was the point of my shutting up the birds in fox-proof houses at night, if foxes were going to come and steal them from under my nose in broad daylight? Of course I realise that foxes have to live, just like the rest of us, and I would happily buy them a chicken from Sainsburys if only they'd leave my birds alone.

CHAPTER 15

The cottage was beginning to take shape now; it's amazing what a difference it makes, once the roof is on. In order to make things easier for the sub-contractors and suppliers, who were having difficulties finding the cottage, the builders erected a large sign at the end of the track. This solved the original problem, only to produce another. Ramblers and sightseers, hitherto unaware that the track led anywhere, started swarming up in their thousands – well, dozens – banging on the door and demanding Dainty Teas, toilet facilities and a Nice Sit-Down. When I pointed out that this was a private residence they were most indignant: 'Well, you shouldn't put up a sign if you don't want people to come.' I suppose there was some sort of logic in this. In any case, they provided so much unintentional humour that it was almost worth the inconvenience.

There was one family I particularly remember – mum, dad and small son. Every question put by the lad was answered solemnly, at great length and with total inaccuracy by dad.

'What's them, dad?' (pointing to the geese).

'Them's swans, son. Very rare specimens, being grey. Swans is usually white. Tropical swans, from America.'

'And what's that?' (indicating Henrietta).

'That's a turkey, lad. Being fattened up for Christmas.'

Lay one tooth on Henrietta, I thought, and I'll rip out your vitals.

They came over and watched me feeding the donkeys. 'What's that you're giving them?'

'Pony nuts,' I replied.

'Oh, I see,' said mum, 'like monkey nuts, only for donkeys.'

'Can I feed them?' asked the boy. 'Can I? Go on, let me, let me feed them, go on.'

'Oh, no,' said mum, shocked. 'Very vicious things, donkeys. They'll bite your fingers off.'

Monty, who was with me, could stand it no longer. Rolling up his trouser-leg, he said, 'Very true – look what they did to me,' and showed them his artificial leg.

Shrieks of horror, followed by hasty retreat of nature-lovers.

Then there were the people who very kindly arranged my life for me. 'What you want to do,' they would announce portentously, 'is put caravans in that field and run it as a caravan-site. Just a waste of space now, with only donkeys on it.' Or 'You ought to get a few more animals and have a wildlife park, like Longleat.' Or 'If I was you, I'd let out that field for pop concerts. You'd make a bomb.' I tried to explain that the whole idea of moving to the country was to get away from it all, not to take it with me. Surprisingly, they chose to take this personally and, muttering something about being able to take a hint, only trying to help, if that's the way you feel about it, we know when we're not wanted, they went off in a huff.

My pet hate was the family who came regularly every Sunday during the summer. They brought a packed lunch which they ate in the garden, very thoughtfully leaving the litter for me to clear away. After banging on the door demanding to use the 'convenience', the children were despatched to play cricket in the paddock. Meanwhile, mum and dad engaged me in an interminable and incredibly boring conversation (well, not really

138

a conversation as I was not allowed to participate, only to listen) about themselves, their children, their holidays, their operations. I would stand there, fermenting with frustration, thinking of all the things I could be doing instead of listening to this rubbish but unable to think of a polite way to break it up. Their parting words were always the same and always brought me to the point of homicidal mania: 'I bet you must be glad to see us, you must get ever so fed up with nobody to talk to all the week.'

On one occasion I was in the donkey-shelter when a pair of middle-aged women came by. Unaware of my presence, they were discussing the cottage in piercing tones: 'I see that cottage has been sold,' said one.

'Yes,' replied the other, 'it's meant to be a secret but I've heard that a famous pop star's bought it.'

'Yes, well,' said her companion, 'they need somewhere out of the way to get away from their fans, don't they?'

I toyed with the idea of throwing myself on their mercy, pleading with them to keep my secret close to their ample bosoms and protect me from the hordes of fans just panting to get their hands on me, but decided against it. One look at me and their illusions would be shattered. Perhaps I should have hung on to my striped hair, as part of the image.

School holidays are a time of great activity, with groups of children walking along the footpath to the woods. They always knock on the door for drinks of water for themselves or their dogs and refills for their water bottles. Really, I thought, after I'd filled about twenty bottles, anyone would think they were going on safari through darkest Africa instead of a stroll through the woods. I got to know some of these youngsters quite well and we developed a mutually satisfactory working relationship – I would give them apples from the garden and they would present me with acorns and conkers in return. Quite a few of the saplings in the garden owe their lives to these swaps.

139

One lad who sticks in my mind is Kevin. When he told me his name it rang a bell, and I asked if by any chance his surname was Timms. 'Yes,' he said, 'why?' Why? Because when I bought the cottage, the inside walls had been completely covered with the most obscene and explicit graffiti, all signed by Kevin Timms. I told him this and watched the colour creeping up his neck and over his face, much to the delight of his companions. 'Oh, that's not me,' he protested. 'That's my cousin. He's called Kevin, too.' Yes, of course!

One really welcome caller was an old gentleman called Fred Hanley. His family had lived in the cottage when he was a boy. This was going back over sixty years to when the cottage formed part of the duke's estate and his father was the duke's gamekeeper. There was no water then, and the loo was a bucket in the barn. Fred and his four brothers used to collect drinking water from the spring two fields away. He told me that as a schoolboy, he had seen a Zeppelin crash in the next field. 'By the time we got there it was blazing like a bonfire. The pilot was dead and ever since, he's haunted the cottage, that being his last resting place.' A ghost! I was thrilled. I've always wanted a ghost of my own. Sadly, he's never materialised since I've been here, but I'm still hoping. I think the cats might have sensed his presence. Sometimes, when we're sitting by the fire at night, they suddenly stiffen, sit bolt upright and stare unblinkingly at absolutely nothing. It can be quite unnerving! London friends often ask me if I'm not nervous, living on my own in the middle of nowhere. When I point out that I'm not living on my own but have a ghost as sitting tenant, they blanch, particularly if they happen to be in the cottage at the time. In fact, I was a lot more nervous living in my London flat than I am here. I would never have dreamed of going to bed without locking and bolting my flat door, but here I never think about it. I've got nothing worth stealing, and I can see cars coming up the track a good ten minutes before they arrive – plenty of time to barricade myself in and call the fuzz.

I have only been really frightened once, and that was soon after I came here. It was late at night and I was in bed in the caravan when I heard somebody coughing just outside. I froze; there's only one person who coughs around here and that's me. I sat up in bed all night, clutching the bedclothes around me and listening to this bronchitic prowler. The phone hadn't yet been installed, which was just as well because when morning came and I was able to muster up the courage to investigate, I discovered that my intruder was one of Joe Sheppey's cows which had broken out of the next field. I hadn't realised that coughing cows sound exactly like coughing people!

Fred told me that the village had changed very little since his day. I could quite believe that. Most of the shops are still family-run and provide the sort of friendly, helpful service you rarely come across in London. I got to know the chemist particularly well because I was always going in for food and medication for the animals. So I suppose it was a natural enough mistake, when I went in for some perfume and asked for a bottle of Intimate, that he should have produced a container of Antimate, a preparation for bitches in heat. He was very apologetic: 'I've got so used to you buying things for the animals, I took it for granted this was what you wanted.' Well, thanks a bundle!

The wine store is another family-run rather unsophisticated business, combining the post office and newsagent as well as grocery. I went in for a bottle of 'cheap red plonk'. The proprietor looked at me over his specs and explained kindly, 'There's no such thing as red blonk. Blonk is white. You mean rooge.' So rooge it was! The high spot of the week is market day, when about fifteen stall-holders take over the market square. Once I discovered the Women's Institute stall, with its marvellous home-made cakes, I gave up baking my own completely

141

– I just couldn't compete. When friends come to tea and ask if these superb creations are home-made, I always say 'yes' with absolutely no compunction. Well, they didn't ask in whose home!

Other callers included people who brought me any injured birds or animals they had come across in the woods – pigeons, crows, starlings, thrushes, squirrels. Charlie was in his seventh heaven and Rufus became more long-suffering than ever. Most of these creatures were just suffering from shock and after a day or two's rest they were ready to go. The injured ones I took to the vet; if their injuries were very severe he gave them an injection to end their suffering. The others he treated and I nursed until they were well enough to be released. The hardest part for me was not to become too involved with these casualties. If they lost their instinctive fear of people their chances of survival in the wild would be virtually nil, so I would be doing them no favours by taming them. Unfortunately, baby creatures who have to be hand-reared invariably attach themselves to whoever cares for them and can't be safely released even if they want to go, which usually they don't. You offer them the freedom of the great wide world and they spend the rest of their lives within a few yards of the only home they've ever known.

Quaggy was one of these, a tiny mallard duckling which some boys had rescued from a dog. He was unharmed but badly shocked. I put him in a hay-lined box with a hot-water bottle and installed him in a cupboard. I knew that if he could get through the night he had a good chance of survival, so next morning I was delighted to see that he was not only alive but noisily demanding his breakfast. I made up a mash for him, which he joyfully splattered all over the floor, the walls and the ceiling. 'You really are a delightful little chap,' I told him. 'Would you like a worm?' He squealed happily, so I went outside and dug one up for him. He was overjoyed with this offering, despatched it in one gulp

and immediately started clamouring for more. And
more. In fact, after that I had difficulty getting him to eat
anything that wasn't a worm and I was soon spending
the best part of each day digging them up for him. The
chickens soon got the message: why dig up worms for
ourselves if we can get Muggins to do it for us? The
minute I picked up the spade they were there, lined up in
strict order of precedence with me right at the bottom of
the pecking order. Mary, being boss lady, sat on the
spade so that she had first option on any worm up-
turned. As she was nearer to the ground than me, she
always beat me to it. Heaven help me, I thought, I never
imagined I'd be reduced to the level of fighting a chicken
over a worm.

Quaggy flourished on his worm diet – very
nourishing, worms are, and full of protein – and soon he
was big enough to face the great outdoors. I wasn't sorry,
because the floor of the caravan was a quagmire and I
spent most of my time scraping bits of his mash off the
walls. He waddled happily along behind me, his bill
about one inch from my heels, quacking joyously. He
was growing fast, but still he followed me everywhere. I
was terrified of accidentally treading on him. When
mating time came around, he suffered an identity crisis.
For one thing, he hadn't seen another duck since
babyhood and probably didn't even know he was a
duck. For another, he identified more closely with me
than with any other creature. As the part of me he had
most contact with was my feet, he fell madly in love with
my wellies. He displayed to them, he courted them and
he did his utmost to mate with them. Callers got used to
seeing me hobbling around with a besotted duck firmly
attached to my foot. Something had to be done. I must
get him a mate, but mallards are wild birds, and I
couldn't see myself duck-napping one from the river. At
last, in desperation, I put a card in the newsagent's
window: 'Lonely Hearts! Lovesick mallard drake
urgently seeks female companion. Object matrimony.' I

didn't get any response from duck-fanciers but I did get
a call from the local newspaper. An enterprising reporter
had seen the card and thought there might be a story in
it. Could he come up with a photographer and interview
Quaggy? Why not, I thought. Anything that brought
Quaggy's plight to the notice of the public at large must
be to the good.

So Quaggy became famous. His picture appeared in
the paper and the world was acquainted with his
problem. I felt a bit mean, exposing him to this sort of
publicity and splashing his personal problems all over
the front page, but I consoled myself with the thought
that my intentions were of the purest and that Quaggy's
friends probably couldn't read anyway. What was more,
the publicity did result in the offer of a lady mallard,
which I accepted gratefully. By this time I knew enough
about the unpredictability of birds and beasts not to get
over-excited about the outcome of my match-making
efforts, which was just as well because Quaggy took one
look at his prospective bride and said, 'No, thank you.'
He was now totally welly-fixated, to the exclusion of

everything else. Maggie, his would-be mate, took her rejection philosophically and immediately set her cap at Josh, to his absolute horror and Mary's outraged indignation.

So where did I go from here? Did I now start advertising for a mate for Maggie? I decided to leave it for a while; Quaggy might come round in time and realise that sex-wise, a real, live mate had considerably more potential than a battered old welly. One thing that saddened me was that while I was still living in the caravan, I couldn't take in any waifs from Jim Barry. His casualties tended to be permanently damaged, or hand-reared foundlings who needed a long-term home, and I just didn't have the facilities for them in the caravan. However, this didn't rule out domestic poultry, so when he turned up late one night with half a dozen hand-reared ducks, I was delighted, particularly when it turned out that there was a drake amongst them – a mate for Maggie, perhaps? It was too late to put them outside so we installed them in the bath overnight. You wouldn't believe the mess six ducks can make in the space of a few hours; it took me three days of solid scrubbing to get the bath into a useable condition again.

Next day I took them out and installed them on the pond while I built a duckery for them. Maggie chummed up with them immediately. Quaggy took one look and went back to his welly.

145

CHAPTER 16

My forebodings about Phil were well founded. He turned out to be the greatest escapologist since Houdini. Even Humphrey was impressed. True, he was smaller than Humphrey so he could squeeze through smaller gaps, but he very kindly enlarged these gaps on his way through so that Humphrey could follow – which, of course, he did. And Sophie followed Humphrey. Meanwhile poor, faithful Simon sat in a crumpled heap at the point of their departure and ate his heart out. Bad as it had been in the old days when he had only had Humphrey to grieve for, it was a thousand times worse now that he had Phil to worry about as well. He wasn't entirely convinced that Sophie was a fit mother – altogether too flighty to his way of thinking, cavorting with Humphrey and neglecting her baby. Much as he loved Humphrey, Simon was by no means blind to his faults. He was disaster-prone. He courted trouble. He trusted strangers! He would lead poor, defenceless little Phil into dangers too dreadful to contemplate. Simon was distraught. He was responsible for Phil and he had failed him. It was all his fault. He buried his head in his hooves and howled silently. I tried to reason with him, pointing out that they had always come back before and that they would come back this time too, but it was no good. If there was one thing Simon's early experiences

had taught him, it was always to expect the worst: 'I know,' he sobbed, 'but this time it's different. This time they've gone for good. I'll never see them again.' And he threw himself into a further paroxysm of grief.

I hated to leave him alone in his misery, but the sooner I rounded up the truants and brought them back, the sooner he'd come out of his slough of despond. Scratching his backside, I explained that I was off to get them, so could he please try not to expire from grief in the meantime. I set off down the path, brooding over the bloody-mindedness of donkeys who seemed to go out of their way to upset my neighbours and at the same time show me up in the worst possible light. As it was, the local farmers found it hard to understand why anybody should want to keep donkeys in the first place. Their attitude towards animals is pretty basic: if you can't milk it, shear it or eat it, why keep it? In other words, there has to be an end-product. (Actually, there is an end-product from donkeys in the form of copious amounts of manure, but Humphrey has assumed sole distribution rights and I am not allowed to remove so much as a shovelful from the paddock.) Understandably, the farmers didn't take kindly to the idea of these parasites invading their barns and tucking into their hay, and I didn't blame them. My mob had plenty of hay of their own and had no need to steal, and it wasn't even as if the forbidden hay was of better quality or a different flavour as I always buy their hay from these very same farmers. 'No,' I told myself, paranoia setting in fast, 'they do it deliberately, to show me up.'

Perhaps the farmer hasn't seen them yet, I thought, not very hopefully, and I can smuggle them out without anybody noticing. Fat chance – they always announced their arrival at the top of their voices and, in case the farmer happened to miss their greeting or was deaf, deposited neat little mounds of donkey manure all over his nice clean yard so that he'd know they'd called. I sought out the farmer and, glueing a sycophantic smile on my face, braced myself for his heart-rending account of how he'd found them in his barn 'tucking into my hay. Poor little buggers, starving they was.' I looked at these fat, overfed, spoilt-rotten little waifs with murder in my heart. Through clenched teeth I hissed, 'Nice donkeys. Mummy's come to take you home,' and moved over to Humphrey to put the head collar on. I knew that if I could lead Humphrey back, the other two would follow. But Humphrey wasn't ready to go home, not while there was still an ounce of melodrama to be milked from the situation. Whimpering piteously, he cowered behind the farmer: 'Don't let her take me. Protect me, poor defenceless little donkey that I am.' Only when he was sure that the farmer had got the message, i.e. that he was not happy at home, that his owner did not understand him, even, Heaven forfend, that she ill-treated him, did he give in and allow me to put the halter on. Carefully adjusting his long-suffering expression, he grudgingly slouched back with me, followed by Phil and Sophie.

I reasoned that it was no use getting screwed up about his behaviour. The first step was to try to understand the basic motivations, so on the trek back I applied a bit of donkey psychology. 'I'm not cross with you, Humphrey,' I lied, 'I just want to understand why you behave like this. Are you insecure? Perhaps you have an inferiority complex and are over-compensating? Were you weaned too early? Were you an Only Donkey, or was there sibling rivalry? Was your mother obsessed with potty training? Tell me all about it.' He turned his soft, mournful eyes on me: 'What are you on about? I do it for fun.'

By the time we were in sight of the paddock, Simon was already pacing backwards and forwards by the gate, braying silently and forming a welcoming committee of one. He ran from one to the other, checking that they were all there. After a session of nose-rubbing and neck-nibbling, he turned to the main item on the agenda – ensuring that Phil had come to no harm. Every hair and

148

whisker was minutely inspected. When he was satisfied that there were no apparent ill effects, he went over him once again, just to make sure. 'You see, Simon,' I said, 'I told you they'd come back, didn't I?' 'Well, yes,' he conceded, 'but there's always next time.' And of course he was absolutely right: there always was next time.

I just hoped and prayed that it wouldn't coincide with Simon's annual inspection by HAPPA, which was due any day now. I didn't see the car coming up the path so the knock on the door caught me completely unawares. Good Heavens, I thought as I opened the door, it's Farmer Giles! The man on the doorstep looked as though he'd been lifted straight from an illustration in a children's book. Sturdy build, ruddy complexion, battered pork-pie hat, old breeches tied at the knees with string: it was all there. All it needs, I thought, is the West Country burr and we've got the complete set. He tipped his hat. 'Mornin', me dear. Name's Bunter. Come to see this owd donkey of ours, if that be all right with you.'

I couldn't believe it. Was he having me on? But no, he was the genuine article all right. It turned out that he really was from the West Country and had spent all his life with horses. He now worked for the Horses & Ponies Protection Association and had come to see that everything was all right with Simon.

My mind worked frantically. I had carried out a donkey stock-check about ten minutes ago and they were all there then, but anything can happen in ten minutes. True, even if the others had gone, Simon would still be there, but I didn't want Mr Bunter to see him prostrate on the ground in a state of collapse. I had visions of myself explaining feebly that he wasn't usually like this, he was just upset because his friends had gone off on a little jaunt. I could imagine the disbelief creeping into his eyes – even to me it sounded pretty weak, so how would it sound to him? And even if Mr Bunter accepted my explanation, what impression would he get of my suitability as a foster-mum if I didn't even know where my charges were from one minute to the next?

Burbling idiotically to cover my apprehension, I took Mr Bunter over to the paddock. Glory be, they were all there. A wave of emotion swept over me: you lovely donkeys, I thought, I knew I could rely on you when it came to the crunch. They lined up by the gate, heads resting on the bar and eyes glowing with expectation. They love visitors because visitors mean apples and carrots and peppermints and other donkey goodies. Smirking sycophantically, they nuzzled Mr Bunter's hand and explored his pockets. Even Simon, whose attitude towards strangers was still a bit unpredictable, offered his rump for a friendly scratch. I felt weak with relief. Mr Bunter inspected them all and pronounced them fit, well and happy. The other donkeys didn't really come under his jurisdiction but I suppose he felt that as long as he was here he might as well give them all the once-over. I was quite pleased; I knew the donkeys were in good shape but it's always nice to have confirmation.

On the way back to the caravan for a cup of tea, Mr Bunter stopped to look at the cottage. This happened to be one of the days when work was actually in progress. He walked all around, making highly critical comments and recommendations at the top of his voice. I was horrified. What if the builders heard him? They would be mortally wounded and show their displeasure in the only way they knew – by downing tools and walking out, probably for ever. I'd have a strike on my hands, the cottage would never be finished and I'd have to spend the rest of my life in the caravan. I had visions of hard-faced pickets marching up and down the path, carrying banners and stopping all traffic to and from the cottage. I would be a prisoner in my own home, having to run the gauntlet of rotten eggs and offensive remarks every time I ventured out to go shopping. I thought of all the hassle – having to deal with deputations, talks round the table. (What table? I hadn't even got a table in the

150

caravan.) I'd have to learn the jargon, like grass roots, parity, negotiation, arbitration, ACAS. It was all too much. I turned an agonised face towards Mr Bunter and begged him to lower his voice, in case the builders heard him. 'What are ye on about, dearie?' he bawled. 'Of course they'll hear me. I wants 'em to. Got to be tough with these layabouts. Mustn't let 'em get the upper hand.' I sent up a silent prayer that the builders should all be struck deaf – just temporarily, of course. 'It's 'cos you're a woman,' Mr Bunter bellowed. 'They takes liberties with a woman. They wouldn't get away with it with me. Oh dearie me, no.' I could well believe it. Pity I hadn't engaged him instead of the chicken-hearted coward I was lumbered with, I thought. I'd have been in the cottage months ago.

It might have been sheer coincidence, but over the next few days the builders really worked with a will and I was much encouraged. Perhaps that's all they needed, I thought, someone to give them a kick up the backside. Unfortunately Mr Bunter's salutary influence didn't last long and soon they were back in the old routine: one hour's work, two hours' break, one week on, two weeks off. I was sorely tempted to ring Mr Bunter with some excuse about Simon not seeming to be up to the mark, just to get him over to administer a booster injection of get-up-and-go to the builders.

CHAPTER 17

A telephone call from Mr Newbury: he has a guinea fowl for me. I run out and tell Henrietta the good news. She raises an exquisite eyebrow. 'One doesn't wish to appear snobbish, but what do we know about this – creature? What is his background? His breeding? Is he suitable? One has to be so careful, you know.' I hadn't thought of that. Still, knowing Mr Newbury, I feel sure that he wouldn't offer me a bird whose blood line was less than impeccable. I tell Henrietta this, but she is unconvinced. She cocks a disparaging eye at Gussie and Griselda: 'Didn't those two oafs come from him?' She's right, of course. I appeal to her better nature. I point out to her that snobbery is the hallmark of the vulgarian, that your dyed-in-the-wool aristocrat takes people for what they are, not for their background and forebears: Noblesse Oblige and all that. She condescendingly acknowledges this and agrees to grant an audience, on the strict understanding that if he proves to be quite impossible I will remove him forthwith. Recognising just how much this concession has cost her, I agree whole-heartedly and set off to collect her consort.

This time I was prepared, so when a disembodied voice asked me what I wanted, I said, 'I want Mr Newbury and I don't want any of your nonsense.' 'Oh, he's not here, he's gone.' Really, this is idiotic, I thought.

'He is expecting me,' I announced grandly. 'Oh no, no, no. Oh no, he's not.' Fortunately Mr Newbury himself appeared at that moment, otherwise I could see myself getting into a nonsensical dialogue with an uppity jay and coming off worst. Mr Newbury took me over to the guinea fowl enclosure and invited me to choose one. They all looked alike to me so I asked him to select a nice, amiable, easy-going bird. After some consideration he made a grab for one and hastily stuffed it into a box. 'There you are,' he said, 'that's the one for you.' I asked him how he could tell. 'Well, that was the easiest one to catch, so it has to be the friendliest.' Logical, I suppose.

I took Percy home and decanted him into a rather classy cage-cum-enclosure that I had knocked up against  the arrival of any unscheduled creature. He wasn't impressed, and bit me viciously to show me just how unimpressed he was. If it came to that, I wasn't all that impressed with him. He marched up and down the enclosure, swearing profusely. I was shocked at his repertoire, which would have done credit to a navvy. I eyed him dubiously. How on earth was I going to explain this lout to Henrietta? I went off to get him some food and water. When I got back, Henrietta was sitting on top of his enclosure, tweeting gently to him. Percy for his part was jumping up and down, hurling abuse at her. 'You really are a miserable little sod,' I told him. Still, be fair, I thought. He's been snatched from the bosom of his family, shoved in a box and transported to a strange environment, only to be shut up in an enclosure. To make it worse, he's surrounded by birds and beasts, all of whom have their freedom. Of course he's fed up. So would you be. But if I let him out before he became acclimatised, he might go astray. I looked at Henrietta, who was clearly taken with this ill-mannered oaf. So much for her elitism, I thought. Perhaps she would keep an eye on him. I decided to chance it.

When I opened the door, Percy hopped out, taking a lump out of my ankle en route. I was getting very disenchanted with him. Completely ignoring Henrietta, he bustled over to the chickens and told them to get lost. They did. Henrietta fussed anxiously behind him, imploring him not to be so hasty; his reply, roughly translated, was 'Piss off.' I was livid. How dare he talk to my lovely Henrietta like that? I was just about to go over and give him his come-uppance when I noticed where he was heading. No need for me to intervene – he was going to do it all by himself. He plodded over to the geese and informed them that their presence was no longer required, so would they kindly leave the premises. Gussie looked at Griselda and Griselda looked at Gussie. Then they both looked at Percy and simultaneously charged. Percy didn't know what hit him. He shot into the air, squawking and cussing like a fish-wife. The geese, delighted with this reaction, laid their necks along the ground and sent him packing. Henrietta, who had been watching this close encounter of the bird kind from a safe distance, flapped miserably behind him, tutting

and fussing and 'I-told-you-so'-ing in her soft little voice. It was too much for Percy: it was bad enough to have been humiliated so dreadfully, without being nagged at by this stuck-up old biddy. He turned round and bit her.

That did it. I grabbed him by his scrawny neck and, resisting the urge to tighten my grip around his wind-pipe, hoisted him onto the gate. I then proceeded to tell him a few facts of life. Briefly summarised, these were to the effect that we are a peaceful community; live and let live is our philosophy. Anybody who is not prepared to accept this might be well advised to consider seeking alternative accommodation. Do I make myself clear? He spat in my face. I had the feeling that my message had not got through.

I tried to make allowances for Percy, but as time went on and he showed no sign of improvement, I was forced to accept the fact that he was just a loud-mouthed, ill-mannered obnoxious little bully. Henrietta was dis-traught; she felt responsible for Percy, Heaven only knows why, and did her utmost to teach him some manners, but it was a lost cause. In the end she decided she could stand it no longer and washed her hands of him. I don't think he even noticed.

During the three months he was here, Percy must have broken all records for making enemies and antagonising people. He hated everyone and everything with a single-minded obsessiveness that was truly awesome. He attacked on sight: ducks, donkeys, cats, geese, wild birds, even me – in fact, especially me. It was like living in a battle area and the rest of us were in a permanent state of fear and exhaustion. When he suddenly dis-appeared one day, I can't pretend that I shed any tears. We all breathed a sigh of relief and picked up the threads of our shattered lives once more. The chickens started laying again, and soon the ducks followed suit. I was much impressed with the ducks because they laid in the duck-house without any of this hide-and-seek nonsense. But compared with the bantams they were very casual about their eggs and simply dropped them like discarded cigarette-ends, losing all interest once they were laid. It never occurred to them to make a nest and actually sit on them. Not that I blamed them: it can't exactly be a bundle of fun sitting on a clutch of hard, knobbly eggs for weeks on end. I really felt sorry for the chickens, for as soon as one batch of babies was hatched and reared, they immediately set to and started another family, so they spent all the glorious summer days just sitting and rearing. It seemed such a waste.

I felt particularly sorry for Clara. As she still insisted on laying on the barn roof, she had nothing to sit on. Nevertheless, she industriously gathered up all the pieces of broken shell, carried them back to the barn roof and settled herself on them. Quite apart from the sheer discomfort of sitting on broken eggshells, it seemed sad that she'd have nothing to show for it, so I made a nest in the barn and put a couple of duck eggs in it, plus a few bits of eggshell so that she'd feel at home. When she came down to feed I grabbed her and installed her on the nest, and there she sat, blissfully unaware that she was going to hatch some very odd-looking chickens indeed.

CHAPTER 18

Who on earth could be knocking on the caravan door at 6.30 on a Sunday morning? An escaped convict seeking refuge? A homicidal maniac lusting for blood? I opened the door and there was the Laughing Policeman. Really, I thought, this is too much. If he's going to start knocking on my door every time he feels like a good laugh, I'm going to have to move. 'Morning, ma'am,' he smirked, fighting to keep a straight face and failing dismally. 'Did you know you've got eight bullocks in your garden?'

Oh, very witty, I thought. I resisted the urge to reply, 'Yes, of course. They're weekend guests,' and said 'No, I didn't know. Tell me about it.'

'Take a look,' he invited.

I pulled on my wellies and went outside and swipe me, there *were* eight bullocks in the garden, standing in a bunch and contentedly eating my saplings. Why on earth can't they eat the nettles and thistles, I thought. I looked at the bearer of good tidings and he looked at me. It was quite obvious that he knew as much about bullocks and the capturing thereof as I did. 'Let's try and get them into the paddock,' I suggested.

'Good idea.'

We continued to stand there, looking at each other, neither of us wanting to put into words what we were both thinking: how? I picked up a stick and he did likewise. Slowly, very slowly, I moved towards the bullocks; he crept along behind me. Behind me! Oh great, I thought. Nothing like a big, burly, intrepid copper to inspire confidence. Feeling like a complete idiot I said 'Shoo. Shoo.' The bullocks looked up from their grazing and eyed me thoughtfully. I moved a bit closer. 'Nice bullocks,' I lied, 'off you go,' and brandished the stick.

To my surprise, they started moving. I was beginning to enjoy this. Sticking my thumb in my belt and metaphorically tipping my stetson, I hollered, 'Move it, you ornery critters! Time to mosey over to the ol' corral.' They moseyed. Flushed with success and crazed with power, I charged after them, yelling, 'Yippee! Heading for the last round-up. Giddi-up, ol' pardners.'

Hearing a stifled snort behind me, I turned round. I had completely forgotten Superman. He was in a bad way; clutching a tree for support, he was trying manfully to control the waves of hysteria that were sweeping over him. 'Oh my Gawd,' he chortled, 'just wait till I tell the boys back at the station about this. They won't believe it.'

'Probably not,' I replied icily. 'Perhaps you'd better bring a video-recorder with you next time you come.' Rotten old spoil-sport, I thought. He hadn't been a lot of help before, but he was absolutely fit for nothing now.

I herded the bullocks decorously over to the paddock gate and ushered them in. The donkeys, who had been observing the rodeo with much interest and amusement from the safety of the paddock, were considerably less entranced when the action moved to their home ground. They huddled together in the far corner of the paddock and watched the cavorting of these intruders with some misgiving. All except Simon: to my utter astonishment, he took off at the gallop and careered after them. He has the most extraordinary run, just like a pantomime horse when he really gets going. I couldn't help it – I just fell about laughing.

Meanwhile the bullocks were getting more and more panicky. When they reached the boundary fence they just kept going, right over the fence. I was astounded – I had no idea cattle could vault fences. Soon there was only one left, either less athletic or more stupid than the rest. Anyway, he just kept going round and round the paddock until I was beginning to feel quite giddy just watching him. When he realised he was the sole survivor, he went berserk and started charging everything in sight: the gate, the fence, the donkeys. I decided I'd better do something before he started on me. There seemed no point in keeping him here when the other seven had gone, so I opened the far gate and out he flew. I had my work cut out keeping Simon from charging out after him. I looked at Simon: 'What was all that about, then?' I asked him. He looked abashed but there was a twinkle in his eye: 'Fun, wasn't it?' You dark horse, I thought.

Now that the crisis was past, Humphrey came swaggering across, full of bravado. 'I would have seen them off,' he said, 'but I had to protect Sophie and Phil.' Yes, I thought – you and Superman both!

It turned out that the bullocks belonged to Graham Christie. He was very apologetic, but I was delighted: it made me feel less guilty about the times my donkeys invaded his premises.

When it was all over, I found I was too restless to settle to anything – there's nothing like a spot of cattle-herding first thing in the morning to throw the whole day out of kilter. It was a beautiful late autumn day and I decided to take an hour or two off just to enjoy it. There wouldn't be many more days like this, and there was already a smell of winter in the air. I pick up a stick and immediately Charlie comes running across. He looks at me reproachfully: 'You didn't tell me you were going walkies.' 'You didn't give me a chance,' I tell him. 'I was just about to call you.' We set off across the paddock towards the woods. Halfway across and Quaggy comes flying over;

159

he circles twice and belly-flops at my feet. He takes up his usual position, one inch from my heels, and, quacking happily, tags along behind. The donkeys tag along behind him. They escort us to the end of the paddock and watch mournfully as we go through the gate into the woods. Humphrey, as leader, gives the signal and they immediately break into the Donkey Serenade. The noise is deafening and I feel sorry for anybody in the vicinity who is trying to enjoy a Sunday morning lie-in. I pick a couple of handfuls of dandelion leaves and, meticulously dividing them into four equal portions, feed them to the donkeys. They demolish them in one gulp and immediately demand more. I spend the next hour gathering dandelion leaves for them and filling my pockets with acorns and pine cones to use for Christmas decorations, bitterly regretting the fact that I didn't bring a bucket with me.

We walk back across the paddock. Henrietta is sitting on the gate, tweeting softly to herself. I search through my pockets and find a raisin, which she accepts gratefully. The chickens and ducks, who never miss a trick, come bustling across, protesting noisily about this discrimination and demanding equal rights. I get the spade and start digging for worms. Josh, ever the gentleman, stands politely aside while his ladies eat their fill. He and Quaggy never get a look-in, so I decide to do it the easy way and give everybody a chance. I fill a container with warm water and detergent and pour it over the ground. Two minutes later, worms are popping up all over the place. The detergent acts as an irritant, forcing the worms to the surface. I feel a bit guilty, because although I don't much like worms, I hate to cause suffering to any creature. Seeing the alacrity with which the birds gobble them up, I console myself with the thought that at least the worms don't suffer for long. I watch them all tucking in. No doubt about it, the best things in life are free: all those lovely, protein-packed worms, and they haven't cost me a penny.

A warning honk in the background brings me down to earth; Gussie and Griselda have just realised that something is going on. They charge over, officiously demanding to know what it's all about and why weren't they invited? We all scatter, leaving them a clear field. They poke at the few remaining worms with their beaks – they don't really like worms, but they can't bear to see anybody else enjoying them.

I pick my way through the long grass, almost stepping on Pudding who is lying sprawled out, waiting for his elevenses to walk by. It never does, of course, but he is nothing if not an optimist. Rufus is curled up on the compost heap, dreaming of the long-gone days when he was young and agile and his joints didn't creak; when he chased butterflies and stalked spiders and his eyes were clear and bright. He opens one eye, sighs asthmatically and goes back to his dream. Clara comes trotting over, closely followed by her two 'babies'. The ducklings are fully grown now and twice as big as she is, but they still follow her around and she still mothers them and fusses over them. And she still goes through agonies every time they go on the pond, convinced that they'll drown. She fidgets on the edge of the pond, imploring them to come out and behave like proper chickens. They can't understand her fear of water: 'Look, mum,' they say, splashing happily. 'It's lovely. Come on in.'

I sit down on a felled tree and think back over the past year and a half. So much has happened, so many memories are permanently etched on my mind. I think about Barney and Jason and experience the old familiar ache that never seems to get any less. I think of the dreadful emptiness of the bathroom without Pooh. In my present mood I even remember Percy with a certain – not fondness exactly, but a kind of respect. It's impossible not to admire a pint-sized bird who is prepared to take on the entire world.

Charlie and I go over to the cottage, followed by Pudding and Min. It is almost finished now and we will be moving in next month, in time for Christmas. 'Won't it be wonderful?' I say to the cats. 'We'll have central heating, hot water, proper rooms with doors on them, knobs on the wall that you just press and the lights come on. There'll be no more filling generators and heaters and Tilley lamps. We'll be able to sit down and watch a TV programme right through. We'll have a proper home again.'

The cats look at me in horror: Live in there? Us? Not on your life! And they shoot off back to the caravan. I have the feeling that this is where we came in.